The Tangram Puzzle Book

A New Approach to the Classic Pieces

The Tangram Puzzle Book

A New Approach to the Classic Pieces

David Goodman

Ilan Garibi

World Scientific

NEW JERSEY · LONDON · SINGAPORE · BEIJING · SHANGHAI · HONG KONG · TAIPEI · CHENNAI · TOKYO

Published by

World Scientific Publishing Co. Pte. Ltd.

5 Toh Tuck Link, Singapore 596224

USA office: 27 Warren Street, Suite 401-402, Hackensack, NJ 07601

UK office: 57 Shelton Street, Covent Garden, London WC2H 9HE

Library of Congress Cataloging-in-Publication Data

Names: Goodman, David, 1960–   author. | Garibi, Ilan, author.
Title: The tangram puzzle book / by David Goodman, Ilan Garibi.
Description: New Jersey : World Scientific, 2018. | Includes bibliographical references.
Identifiers: LCCN 2017045461| ISBN 9789813234000 (h : alk. paper) |
    ISBN 9789813235205 (pbk. : alk. paper)
Subjects: LCSH: Tangrams. | Mathematical recreations.
Classification: GV1507.T3 G37 2018 |
LC record available at https://lccn.loc.gov/2017045461

British Library Cataloguing-in-Publication Data
A catalogue record for this book is available from the British Library.

For any available supplementary material, please visit
http://www.worldscientific.com/worldscibooks/10.1142/10819#t=suppl

Printed in Singapore

Preface

The Tangram is an old Chinese puzzle which consists of seven pieces. Each piece is made up of one or more right-angled isosceles triangles. In total there are 16 triangles and the objective is to assemble these pieces into a given shape.

Over the years, thousands of puzzles have been created with those seven pieces. Most of them require us to figure out the arrangement of the pieces just by looking at the silhouette of the shape.

This fresh and original book presents a collection of different types of puzzles. The puzzles we present here use the pieces as building units only, and present a variety of challenges, from all fields of recreational mathematics. For example, you will find symmetry puzzles, cover-up puzzles and even a Poker-related puzzle.

In some puzzles we may not use all seven pieces of the Tangram set, while in other puzzles, we may use pieces from *two or more* Tangram sets.

In a twist to the classic Tangram puzzles, for most of the puzzles in this book we will not allow connection between pieces, unless they share a common edge; a point-to-point connection is forbidden. With this slight change in rules, we opened up a whole new universe of puzzles and challenges!

Most of the puzzles presented here are new and original. While some are based on classic puzzles, there are also new and modified ones.

We hope you like the variety. If you do, we would love to hear from you and get your feedback. In addition, if you have queries, new solutions or new Tangram puzzles, please write to us at thetangrampuzzlebook@gmail.com.

Thanks

This book could not have been written without the generous help of many people.

Our team of testers: Dor Tietz; Salomon (Sallie) Hollander; Gadi Vishne and Yael Meron. Their help made the puzzles in this book much better.

Also, there are many puzzlers around the world who contributed their own puzzles to this book, or pointed us in the direction of new genres or celebrated "oldies". To all those from far and wide, a big Thank You.

Contents

Introduction to Tangram

What is Tangram?

The Tangram is an ancient Chinese dissection puzzle. Its building blocks are seven flat pieces, called *Tans*. The objective is simple — given the outline of a shape, determine the arrangement and orientation of the seven pieces of the Tangram.

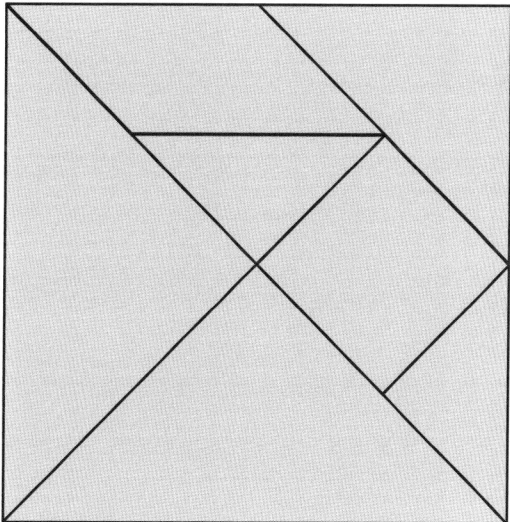

The arrangement of the seven puzzle pieces to form the
classic shape of a square.

It is said to be invented in China during the Song Dynasty, and then introduced to Europe by traders in the early 19th century. It became very popular in Europe and in a short time went on to become one of the most popular dissection puzzles in the world.

The seven pieces of a Tangram set are all made up of the same basic unit — a right-angled isosceles triangle. The seven pieces are:

- two large right-angled triangles, made up of four units each;
- one medium right-angled triangle, made up of two units;
- two small right-angled triangles, made up of a single unit each;
- one square, made up of two units;
- a parallelogram, made up of two units.

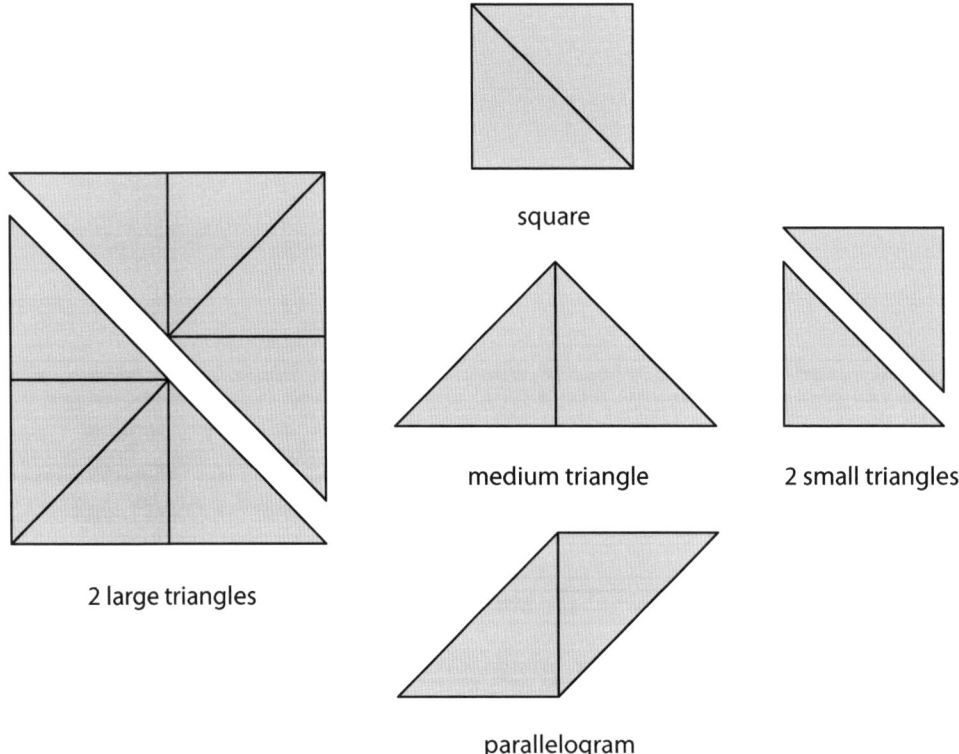

square

medium triangle

2 small triangles

2 large triangles

parallelogram

Of these seven pieces, the parallelogram is unique in that it has no reflection symmetry but only rotational symmetry. If you want its mirror image, you have to flip it over. It is the only piece that may need to be flipped when forming certain shapes.

The Puzzle and How to Solve It

The silhouette given is usually a continuous shape, but there may be some shapes with empty spaces inside.

Look at the shape of the lady given below. It is easy to locate the big pieces first, which are the hat and the skirt. Next, the hands (parallelogram) are evident, as are the face (small triangle) and the legs (square). Lastly, the medium and the small triangles make up the body.

Classic Examples

Here are some classic basic silhouettes.

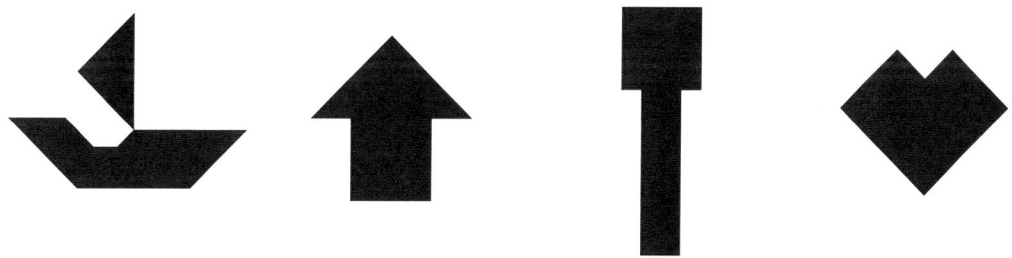

Solutions

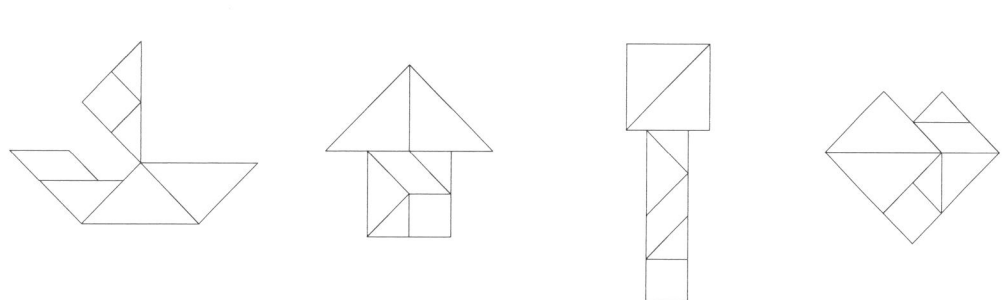

The Game Variation

With the years came the Tangram Paradox, demonstrated here with the Magic Dice Cup paradox, taken from Sam Loyd's *The Eighth Book of Tan* (1903).

All three cups use the same pieces of the Tangram set. Yet, they are not the same. The middle one contains an opening at the top, while the one on the right has an empty space.

Solutions

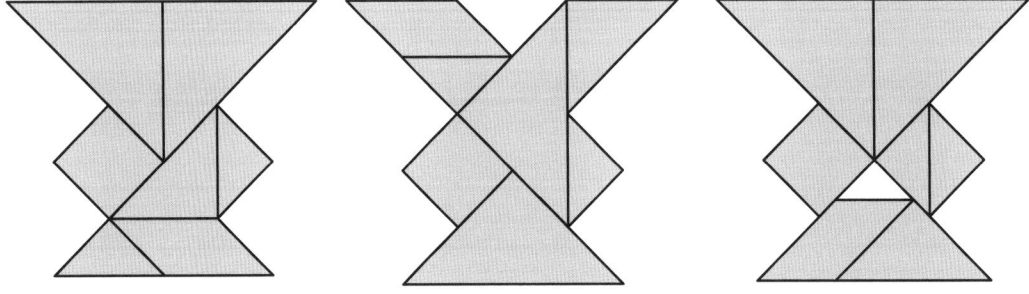

The Pieces Variation

Over time, people came up with different shapes for the pieces. There is a round-shaped Tangram, an oval-shaped one, an eight-piece Tangram set and many more other variations.

heart-shaped tangram

egg-shaped tangram

Glossary of Terms

1. **A piece** — any one of the seven pieces of the Tangram set.

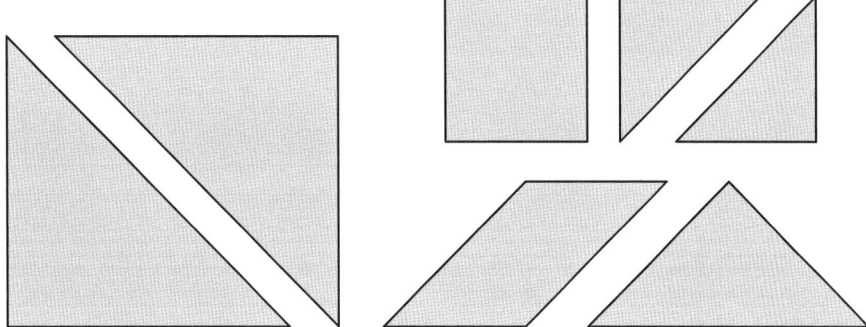

2. **A unit** — the smallest right-angled isosceles triangle. All the bigger pieces are built from this unit. The square, the medium triangle and the parallelogram are each made up of two units. The large triangle is made up of four units, and the whole set is made up of 16 units.

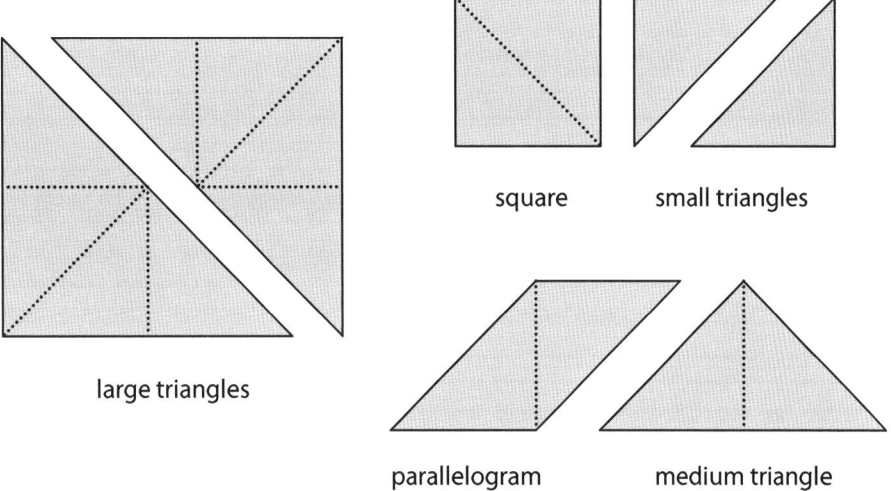

square   small triangles

large triangles

parallelogram   medium triangle

3. **A part** — a shape made up of two or more pieces.

4. **Identical parts** — parts that are the same in shape and size.

5. **Symmetrical shape** — a shape made up of two halves that are exactly the same in shape and size. The symmetrical shape can be translational symmetric; reflectional symmetric (mirror image) or rotational symmetric (symmetric with respect to a point).

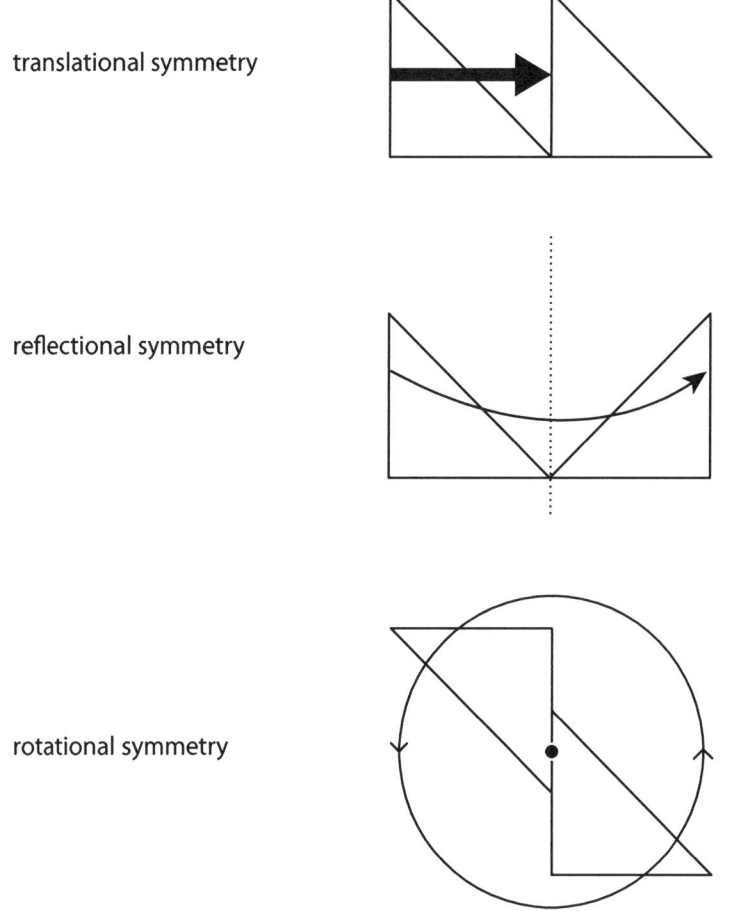

6. **Convex shape** — a shape or polygon in which all its interior angles are less than 180°.

7. **Triboloes** — shapes that are created with three right-angled isosceles triangles.

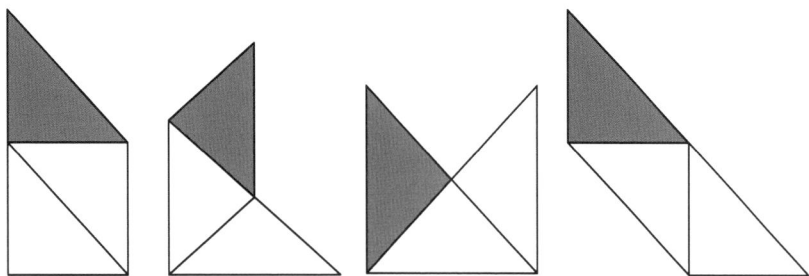

There are altogether 4 triboloes, as shown here.

8. **Tetraboloes** — shapes that are created with four right-angled isosceles triangles.

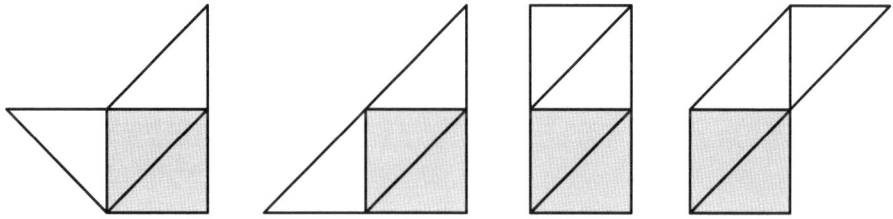

There are altogether 14 tetraboloes; these are given in the Appendix.

9. **Pentaboloes** — shapes that are created with five right-angled isosceles triangles.

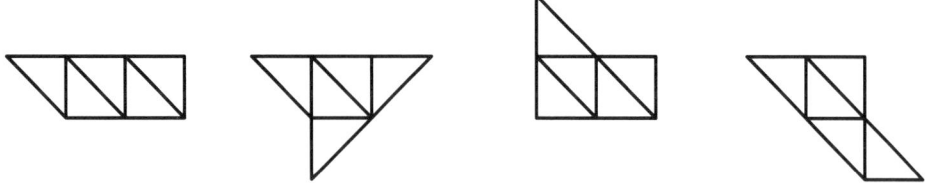

There are altogether 30 pentaboloes; these are given in the Appendix.

10. **Pentominoes** — shapes that are created with five squares. Each square must touch another along its full edge only.

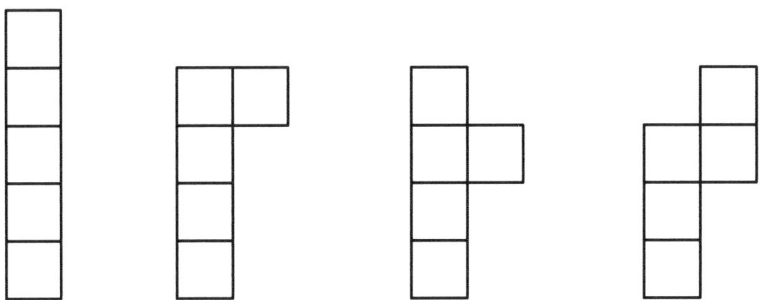

There are altogether 12 pentominoes; these are given in the Appendix.

How to Use this Book

Each puzzle in this book is independent, and you can solve them in any order you wish.

The standard rules when solving Tangram puzzles apply:

1. All pieces must lie on a flat surface. You may not place a piece on top of another.

2. The pieces have to form a continuous shape, with every piece touching another.

In this book, we introduce some additional rules and restrictions to make the puzzles more challenging:

3. Unless otherwise stated, the pieces must touch and align fully along their edges (from endpoint to endpoint) if the edges are of the same size. If the pieces are of different sizes, then their units must align. Below are some examples of proper and improper alignment of pieces.

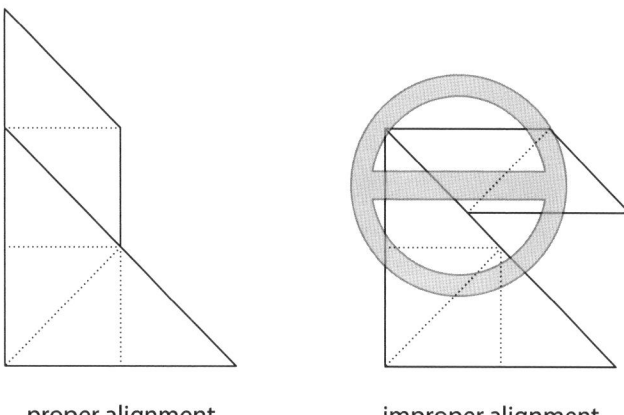

proper alignment   improper alignment

When edges are not of the same size, then they must be aligned along their units.

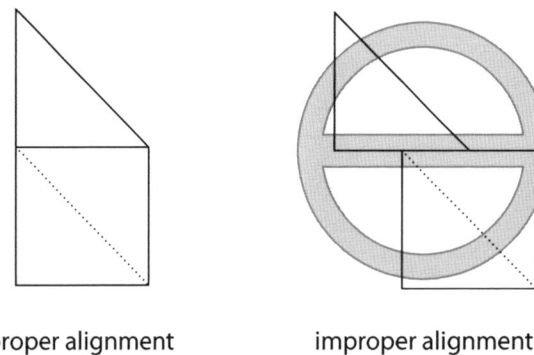

proper alignment   improper alignment

When edges are of the same size, they must align fully.

4. When pieces are presented in a puzzle as a part, do not separate them! Instead, use them together, as if they are a single, bigger piece.

5. For all division puzzles, a piece can be divided into its units, unless otherwise stated. But a unit cannot be divided!

Divide the square into two identical parts:

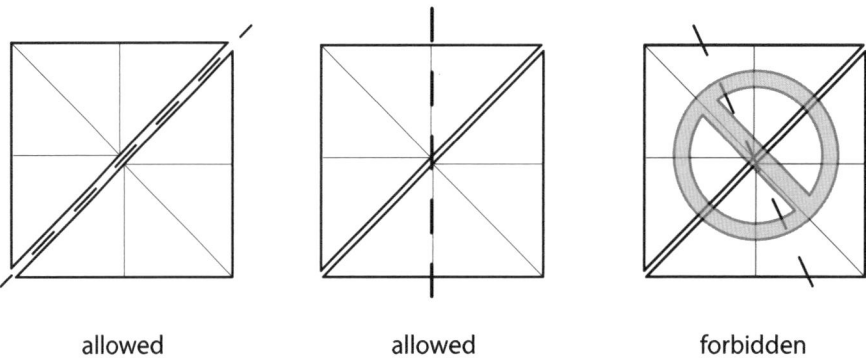

allowed   allowed   forbidden

6. Two shapes are said to have the **same area**, if the number of units in both shapes is the same, regardless of the arrangement of the pieces.

7. In some of the puzzles, the diagrams presented are just generic examples of the solutions we seek. We do not present the actual outline or dimensions so as not to give away the solution.

In Chapter 4, we may suspend some of these rules to create some innovative and interesting puzzles!

Puzzle Types

1. **Paradox Puzzles** — present two shapes that are almost identical. There are two versions:

   (a) both shapes use all seven pieces of the Tangram set
   (b) one shape uses only six pieces while the other uses all seven pieces.

2. **Maximum [Polygon] Puzzles** — make the **maximum** number of a certain polygon (a triangle, a square, etc.) by using as many of the seven Tangram pieces as possible.

3. **Complete [Polygon] Puzzles** — make the **maximum** number of a certain polygon by using **all** seven Tangram pieces.

4. **Negative Puzzles** — in this puzzle type, you focus on the white space in a given black frame. Locate a single piece (or two pieces) such that it reveals the location and orientation of the remaining pieces that make up the white space. Here is an example:

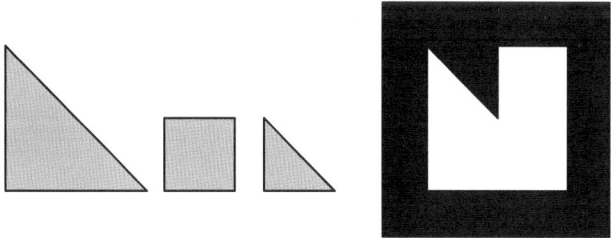

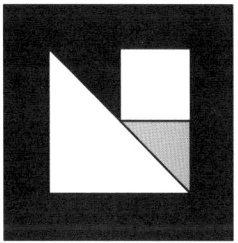

5. **Symmetrical Puzzles** — use the pieces to create a symmetrical shape. The actual shape is unknown, and is the essence of the puzzle.

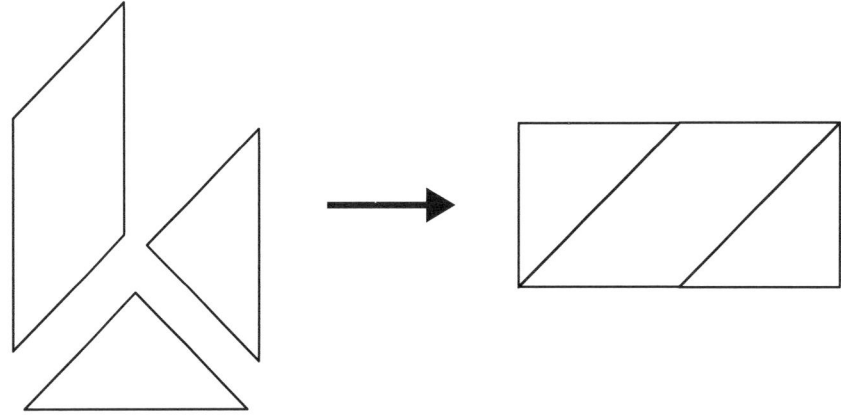

Example of a symmetrical puzzle.

Chapter 1
Tangram Set Puzzles

For this chapter you need a single, complete set of the Tangram.

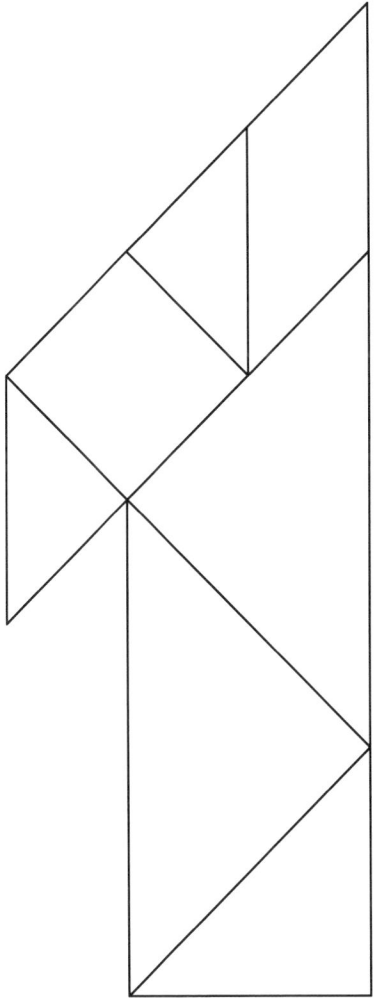

1.1.   *The House*

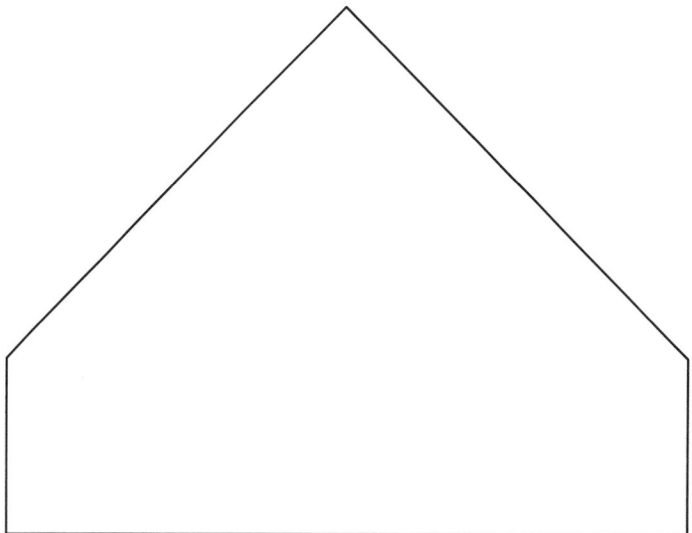

Arrange the seven pieces of the Tangram to make the shape above.

Find at least three different solutions:

(a)   a solution where you can divide the shape into two parts having the same area by one straight cut along the lines of the pieces.

(b)   a solution where, by "folding" the five small pieces onto the two big ones, the small pieces will cover the big ones, showing that their areas are equal.

(c)   a solution where the two large triangles are not connected by their edges.

1.2. *The Rocket*

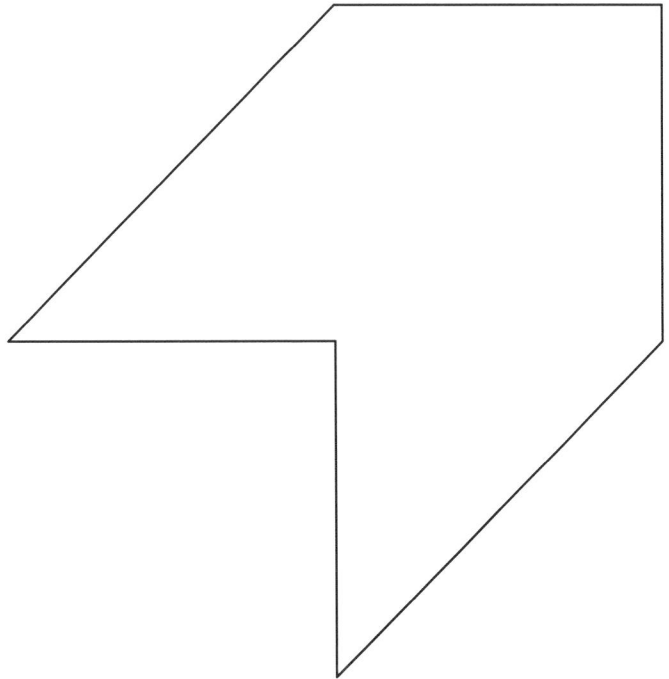

Step A: Create the shape above, using all the pieces of a Tangram set.

Step B: Divide this shape into four identical (in size and shape) parts, each part with a size of four units.

There are at least three solutions.

1.3. *Maximum Squares*

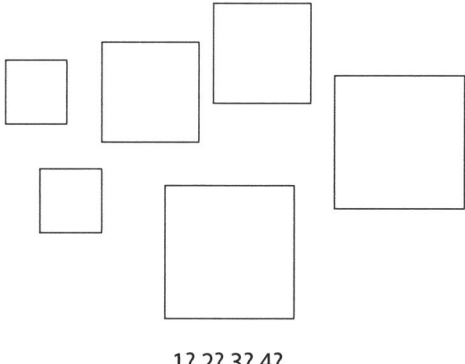

1? 2? 3? 4?

Make as many squares as you can using the seven pieces of a single Tangram set.

Use as many of the seven pieces as possible, but each piece can be used only once.

1.4. *The Hammer Head*

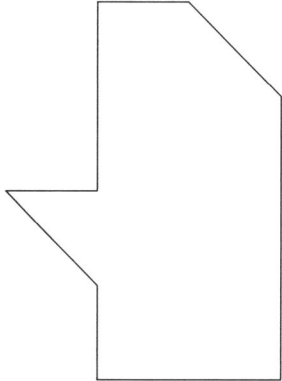

Step A: Create the shape above using all seven pieces of a Tangram set.
Step B: Divide this shape into four identical (in size and shape) parts, each part with the size of four units.

There are at least two solutions.

1.5. *The Spinning Top Paradox*

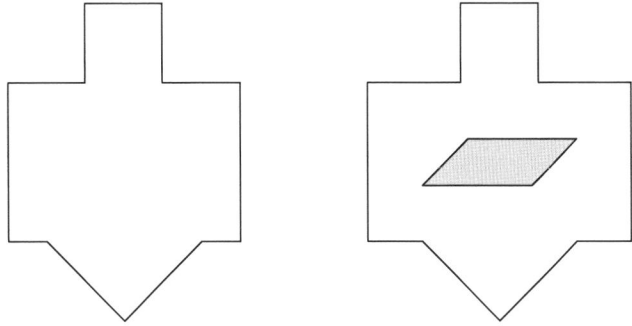

Both of these spinning tops are each made up of a single Tangram set. Recreate the two shapes.

Note: In this puzzle, the pieces need not align fully along their edges.

1.6. *Parallelograms*

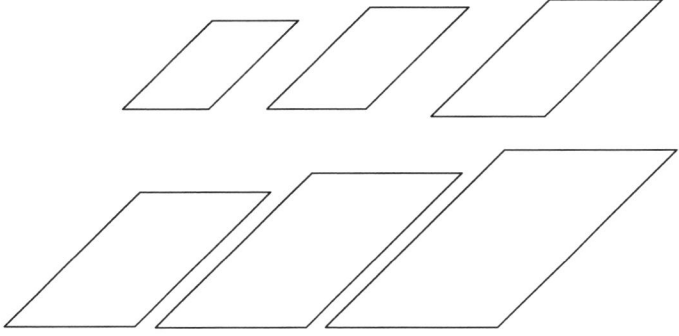

Make six parallelograms using just two, then three, then four, then five, then six and finally all seven pieces of a full Tangram set.

Note: For this puzzle, all the parallelograms must have two 45° angles.

1.7.   *The Wedge*

Step A: Create the shape above using all seven pieces of a Tangram set.

Step B: Divide this shape into four identical (in size and shape) parts, each part with the size of four units.

There are at least three solutions.

1.8.   *Complete Triangles*

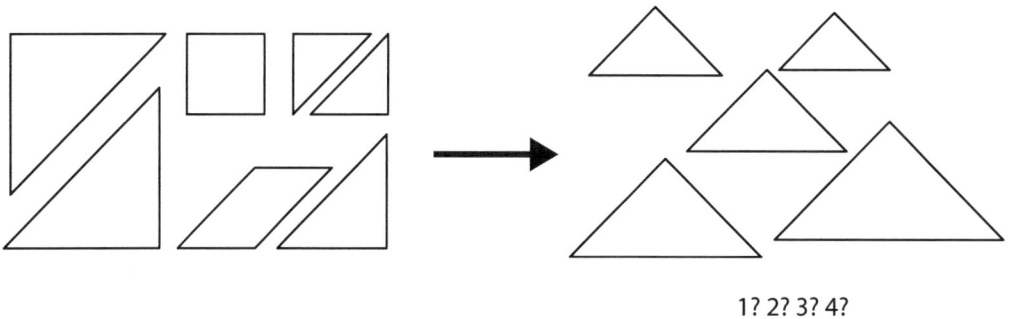

1? 2? 3? 4?

Make as many triangles as possible using the seven pieces of a single Tangram set.

You must use all seven pieces, but each piece can be used only once.

1.9.   *The Mind Reader*

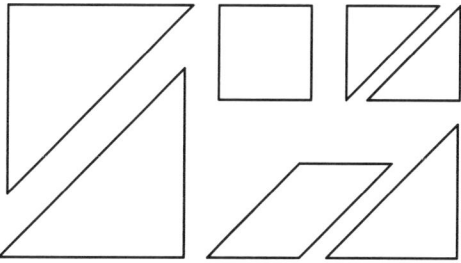

Take out any two-unit piece from a full Tangram set.

Divide the rest into two groups, with each group consisting of three pieces with a total of seven units.

Create the <u>same</u> symmetrical shape with each of the two groups of three pieces.

1.10.   *Complete Trapeziums*

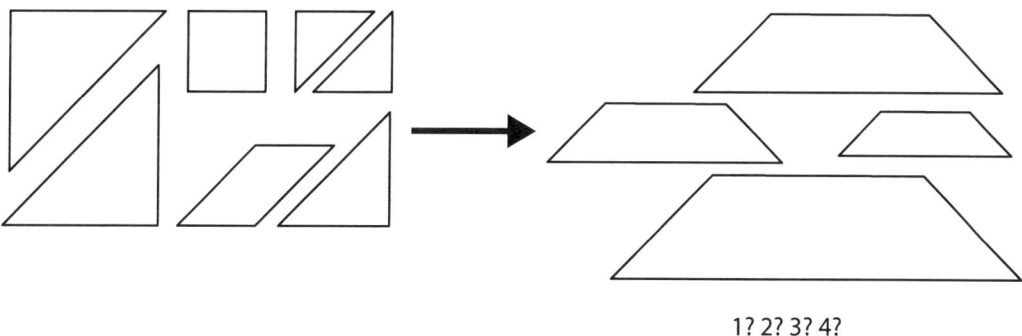

1? 2? 3? 4?

Make as many trapeziums as possible using all seven pieces of a single Tangram set.

Each piece can be used only once.

Note: For this puzzle, all trapeziums must have at least a single 45° angle.

1.11. *The Triangle Dues*

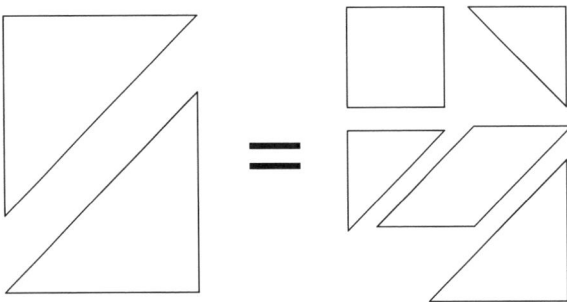

Divide the seven Tangram pieces into two groups: one consisting of the two large triangles, and the other consisting of the remaining five pieces.

Using the first group of two large triangles, create as many shapes as possible. For this puzzle, the two large triangles need not align fully along their edges; they can be aligned along a unit.

For each shape created, cover the two-triangle shape completely with the other group of five pieces.

Find at least five solutions.

1.12. *Complete Parallelograms*

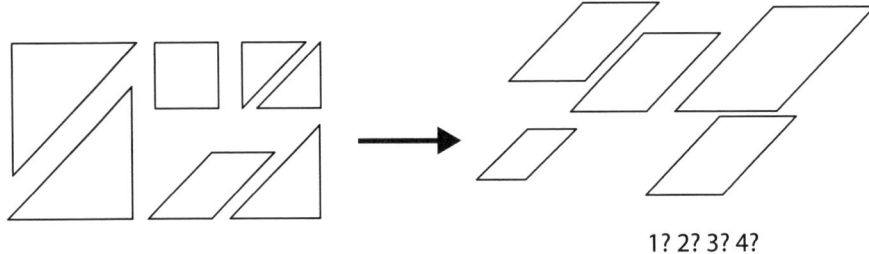

1? 2? 3? 4?

Make as many parallelograms as possible using all seven pieces of a single Tangram set. Each piece can be used only once.

Note: For this puzzle, all the parallelograms must have two 45° angles.

1.13.   *The Trapezium Paradox*

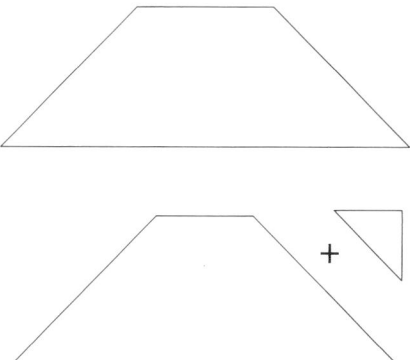

Make a trapezium using the seven pieces of a single Tangram set. Then take out a small triangle and make another trapezium.

1.14.   *The Convex Sixteen*

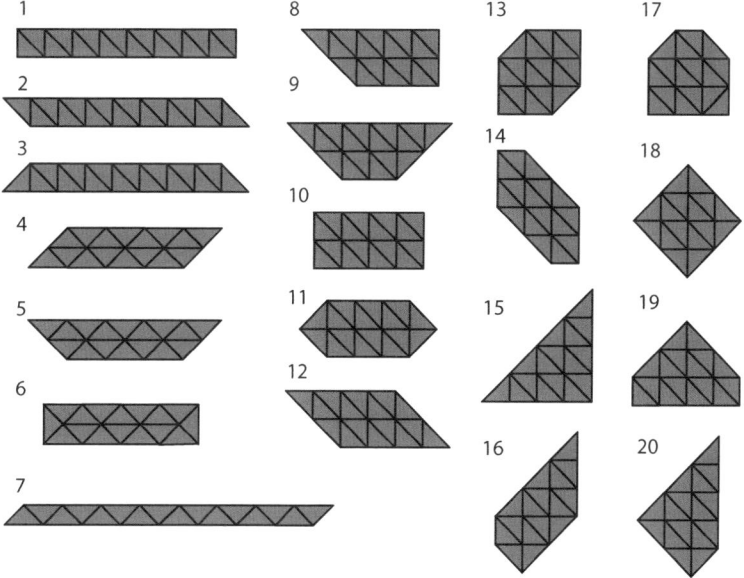

By using 16 right-angled isosceles triangles, you can make 20 convex shapes, as shown above.

Which of them can be created using the seven pieces of a single Tangram set?

1.15.   *The Arrow*

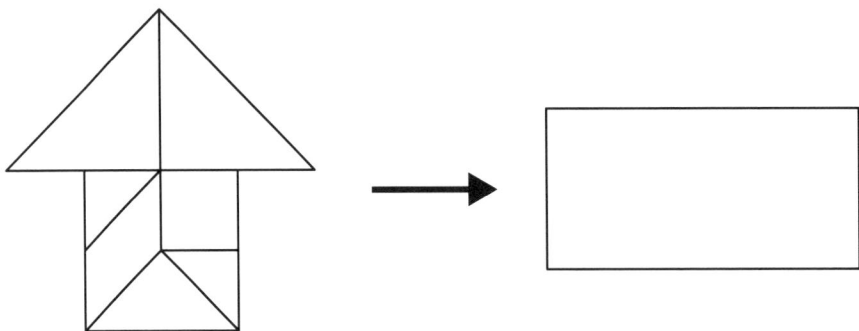

Step A: Divide the arrow shown above into three parts, with each part having a different shape.

Step B: Rearrange these three parts to get a rectangle.

1.16.   *The Bump*

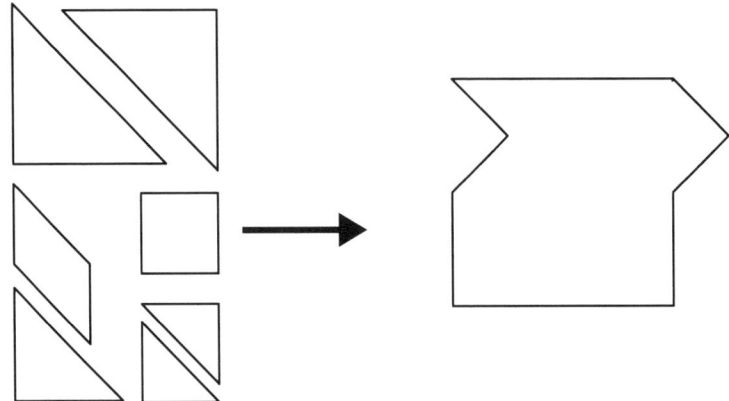

Step A: Create the shape above using a full Tangram set.

Step B: Divide the shape into four identical parts.

Step C: Cut the shape into five identical parts without following the edges of the units. (This is a trick puzzle.)

This is the Tangram variation of a puzzle by **L. Vosburg Lyons**, USA.

1.17.   *The Negative Puzzle*

The white space inside the black frame is made up of the seven pieces of a Tangram set. Locate just two of the pieces in the white space so as to reveal the arrangement of the rest of the pieces.

Solutions

1.1. *The House*

(a) "Cutting" along the dotted line divides the shape into two parts having the same area.

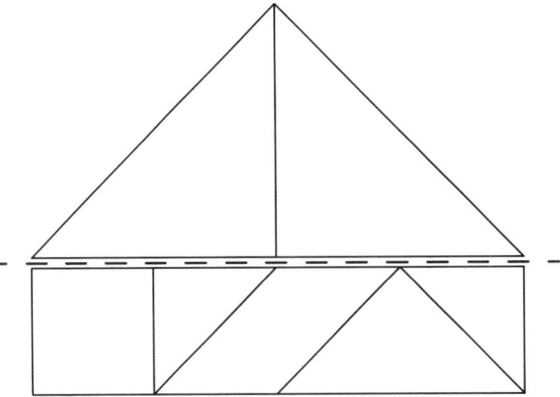

(b) "Folding" the five small pieces as shown below will cover the two big pieces.

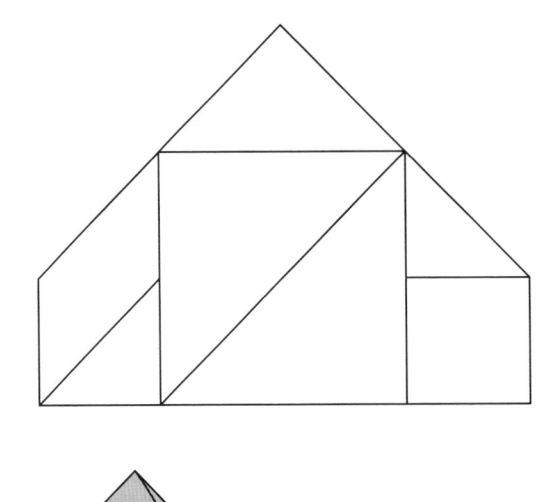

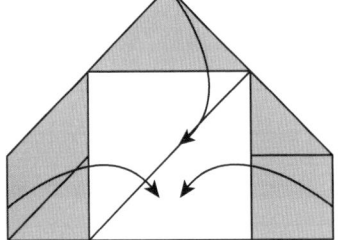 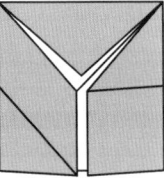

(c)   The two large triangles are not connected by their edges.

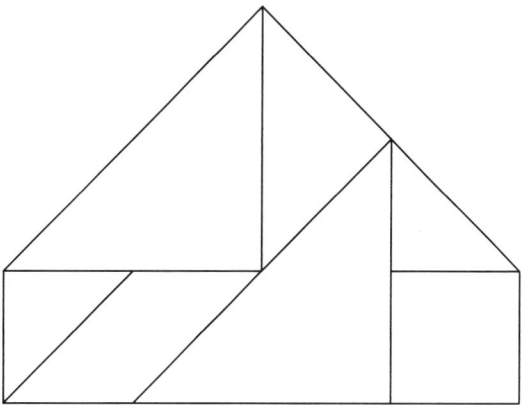

1.2.   *The Rocket*

Step A

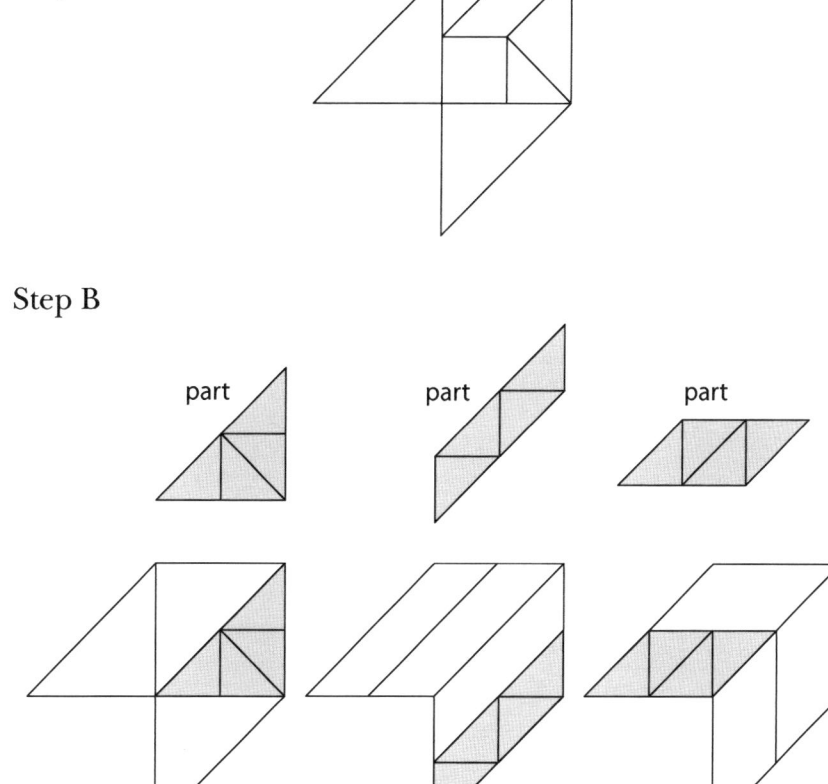

Step B

1.3. *Maximum Squares*

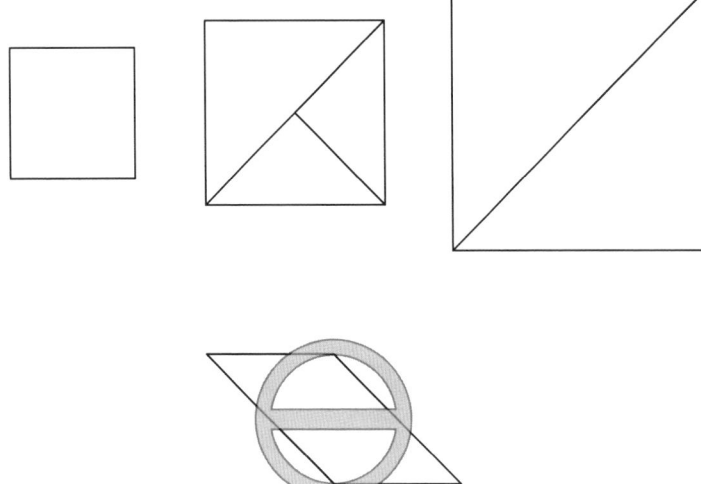

1.4. *The Hammer Head*

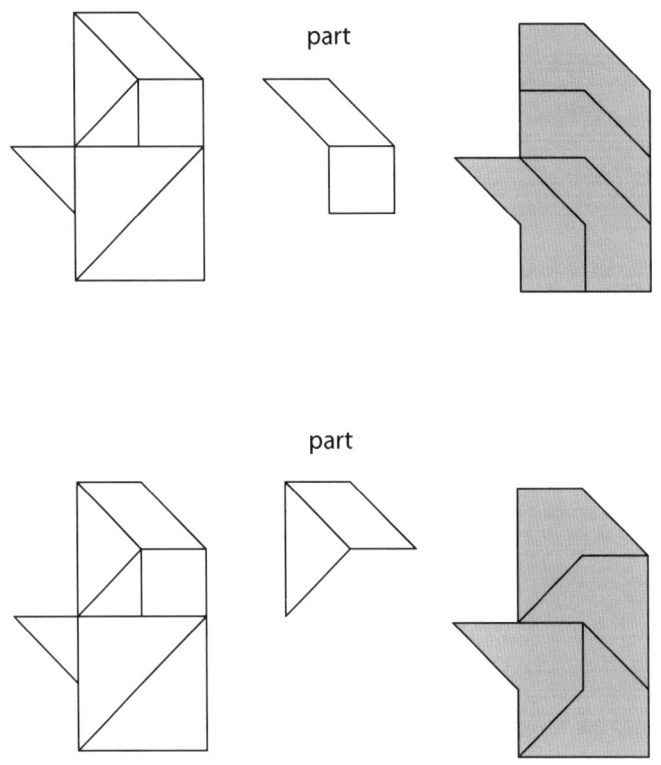

part

part

1.5. *The Spinning Top Paradox*

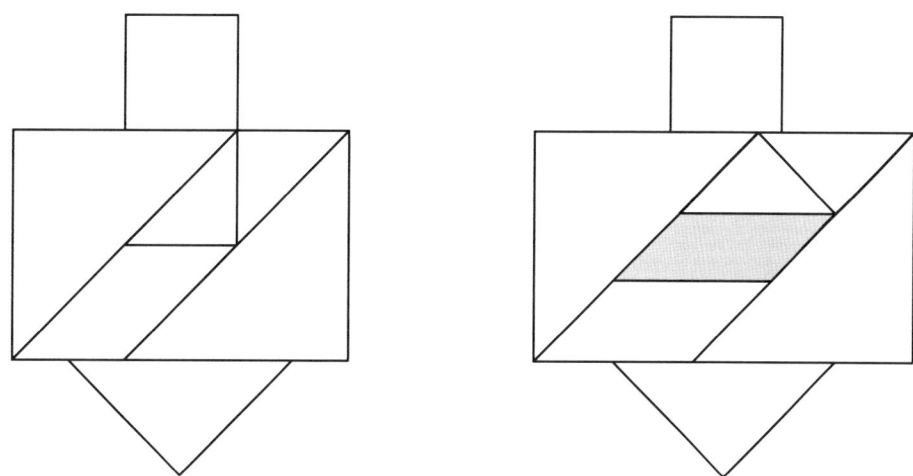

1.6. *Parallelograms*

1.7.   *The Wedge*

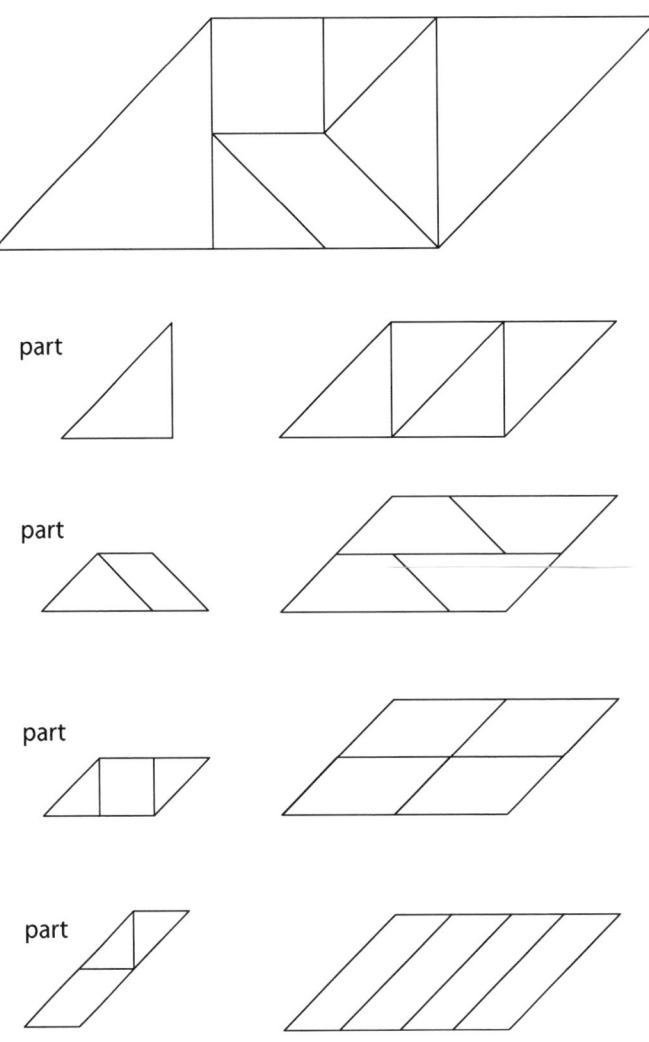

part

part

part

part

1.8.   *Complete Triangles*

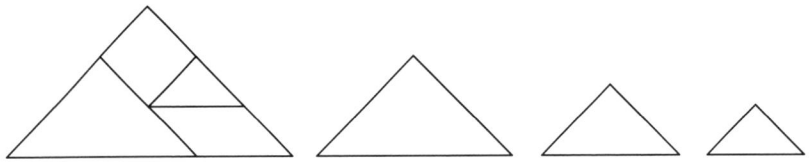

1.9.   *The Mind Reader*

1.10.   *Complete Trapeziums*

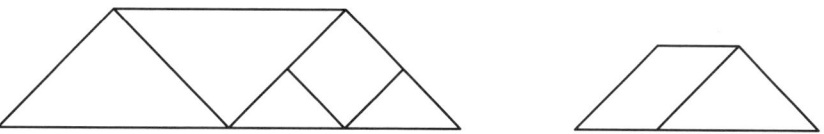

Another option

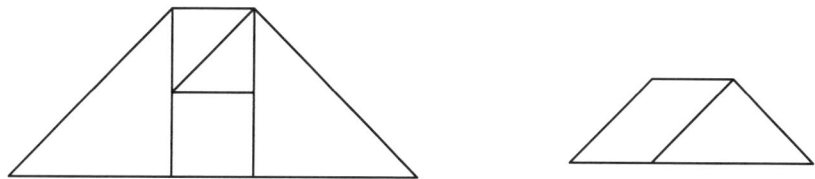

1.11. *The Triangle Dues*

There are six possible shapes, but only five can be fully covered by the other pieces.

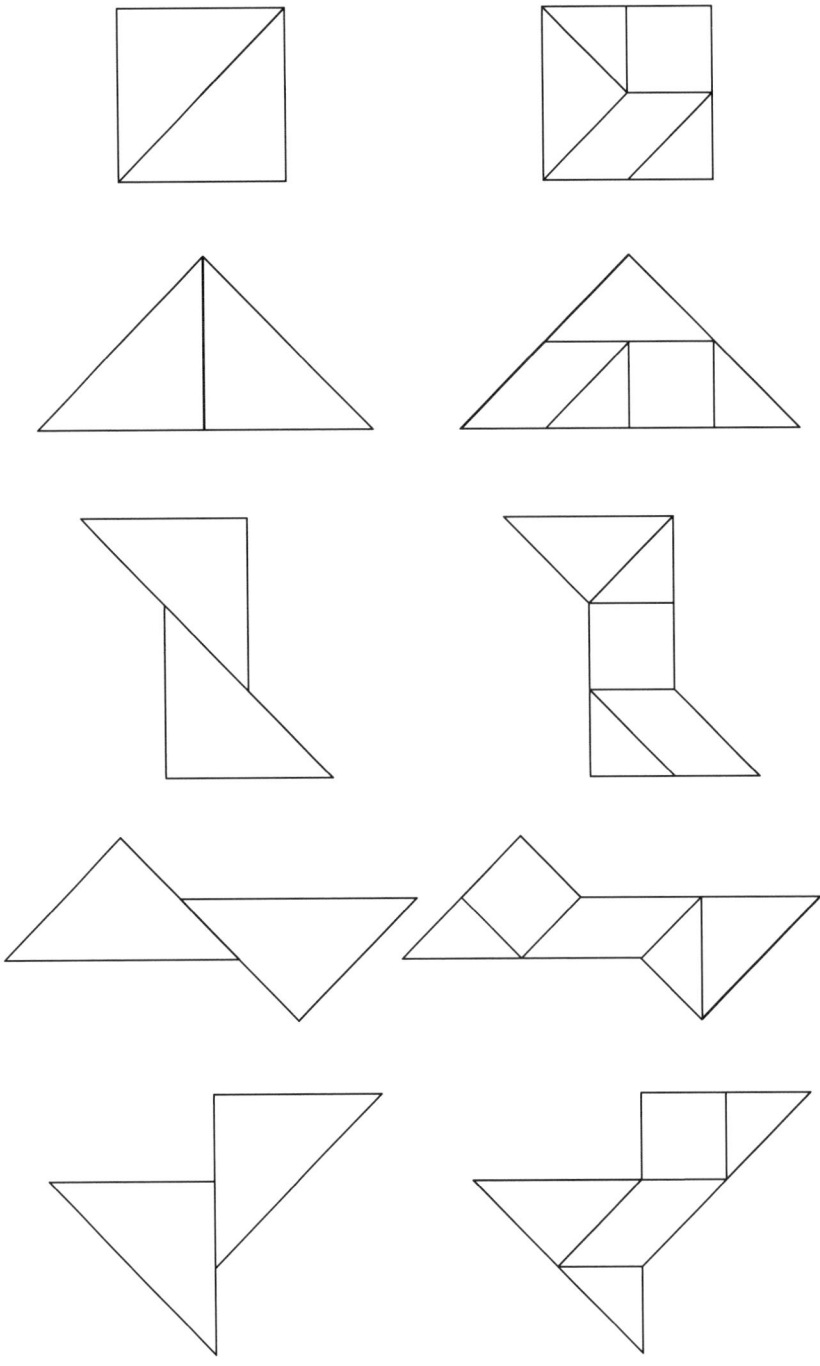

The following shape cannot be covered:

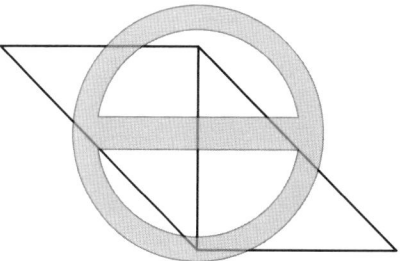

1.12. *Complete Parallelograms*

1.13. *The Trapezium Paradox*

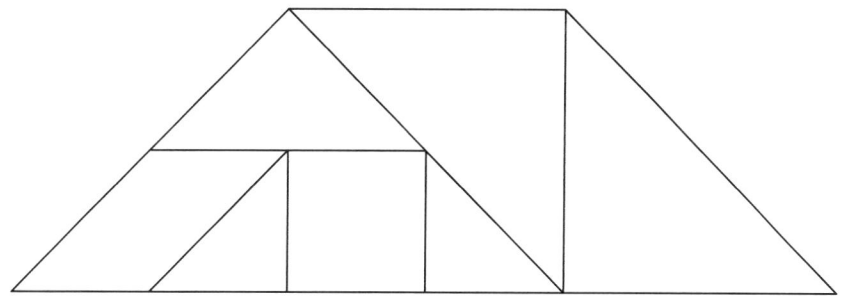

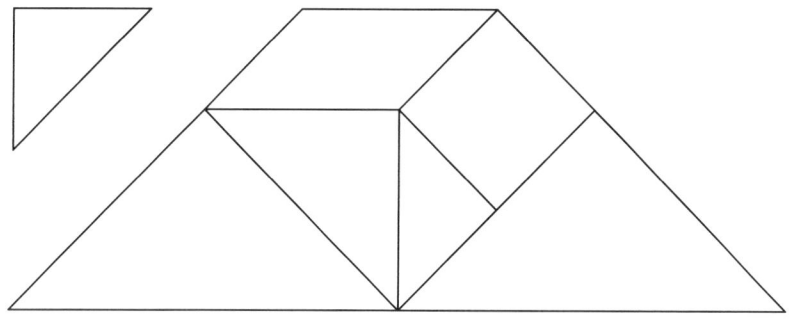

1.14. *The Convex Sixteen*

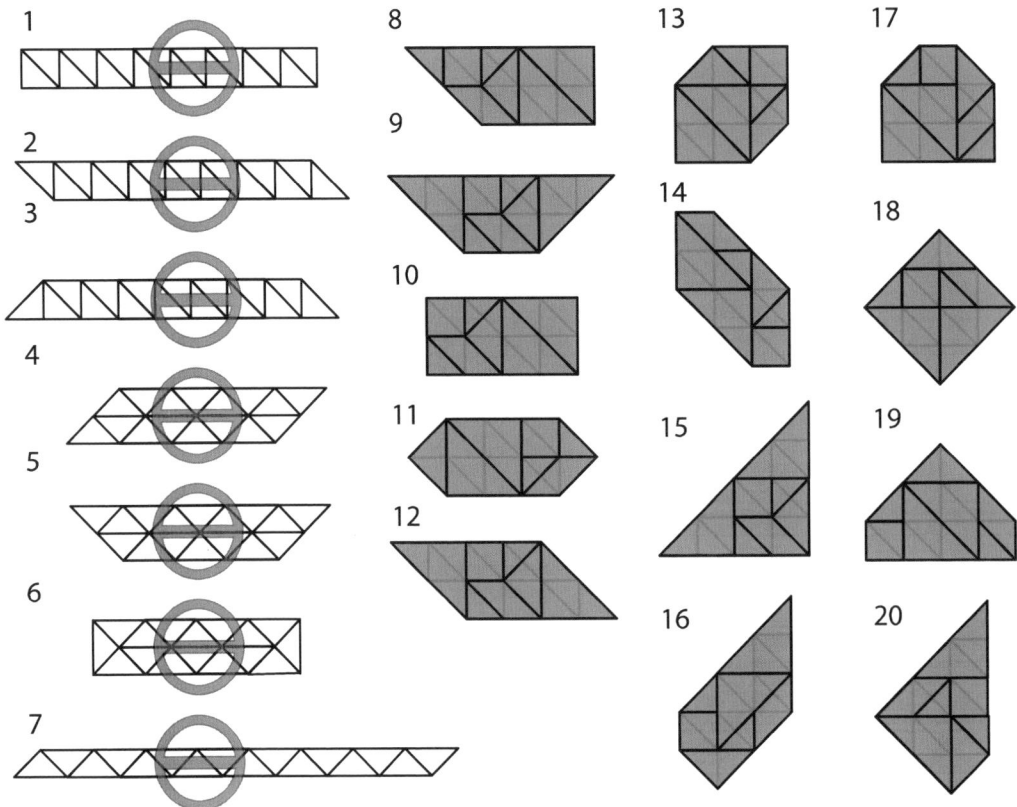

Shapes 1 to 7 cannot be created with a complete Tangram set.

1.15. *The Arrow*

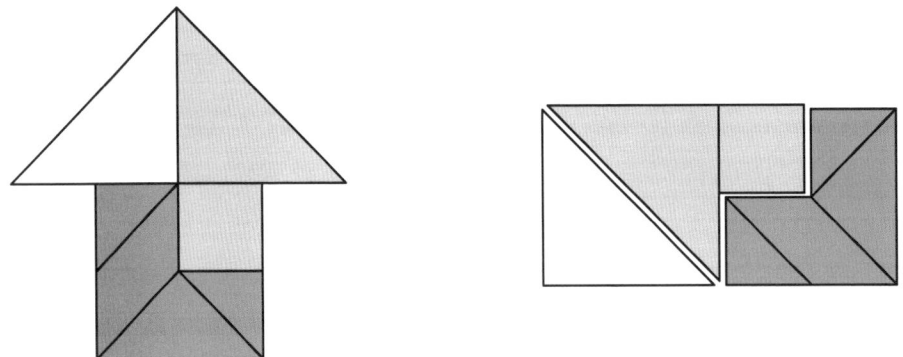

1.16. *The Bump*

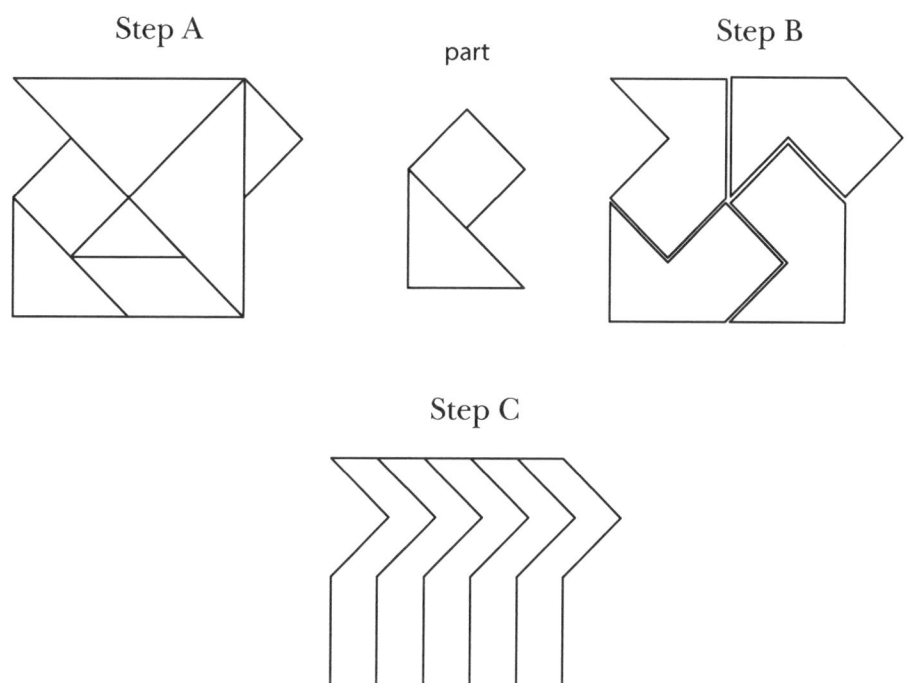

Step A

part

Step B

Step C

1.17. *The Negative Puzzle*

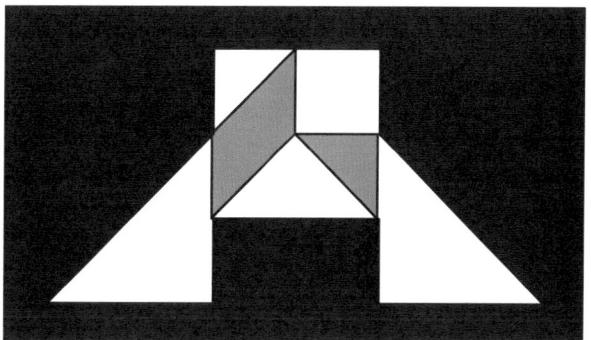

Chapter 2

Incomplete Tangram Set Puzzles

In this chapter we do not use all the pieces of a Tangram set. We indicate this either as a set minus the missing parts, or we mark the pieces that are not in use with a No Entry sign, as shown below.

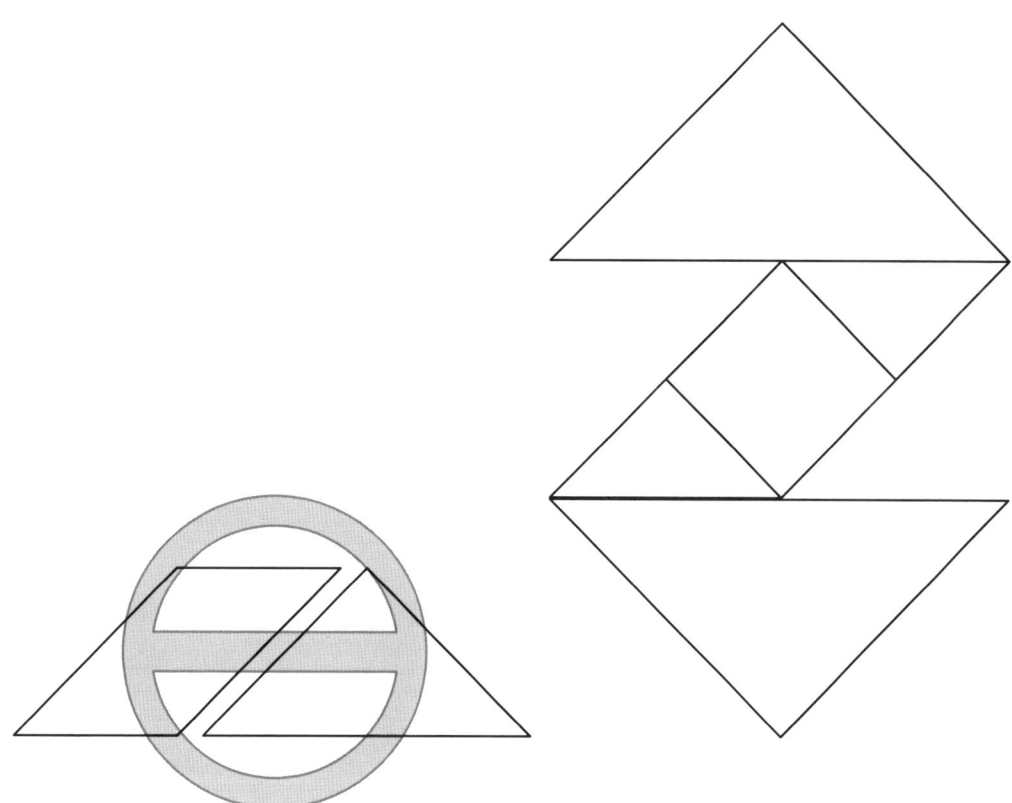

2.1.   *Six-Piece Rectangle*

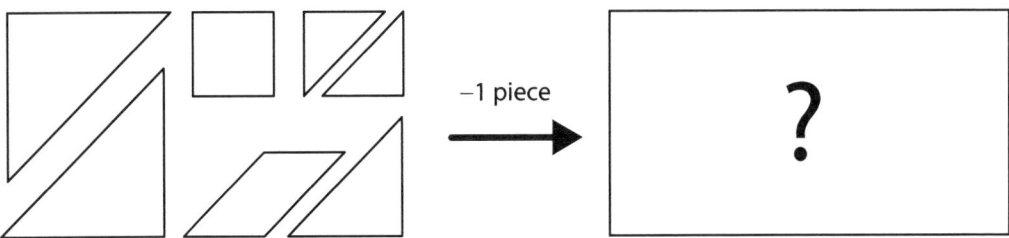

Choose six pieces, and arrange them into a rectangle.

2.2.   *The Fifth Division — Diamond*

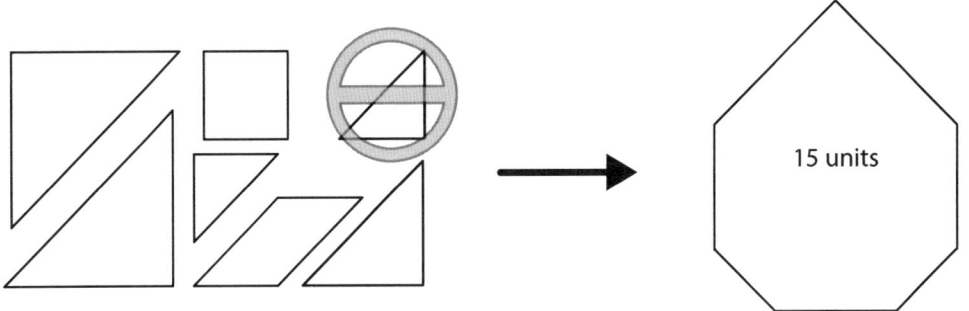

Step A: By using all but the small triangle, create the symmetrical shape above.

Step B: Divide this shape into five identical parts.

2.3. *The Butterfly*

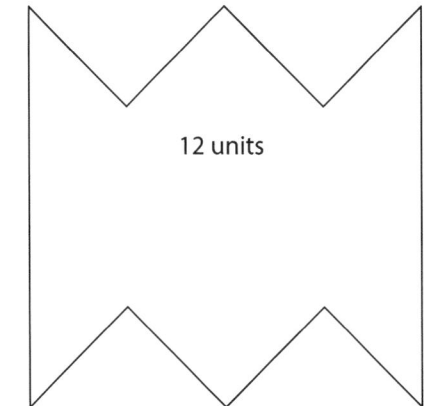

12 units

Step A: Create the shape above using pieces from a single Tangram set with a total of 12 units.

Step B: Divide it into four identical parts.

2.4. *The Third Division*

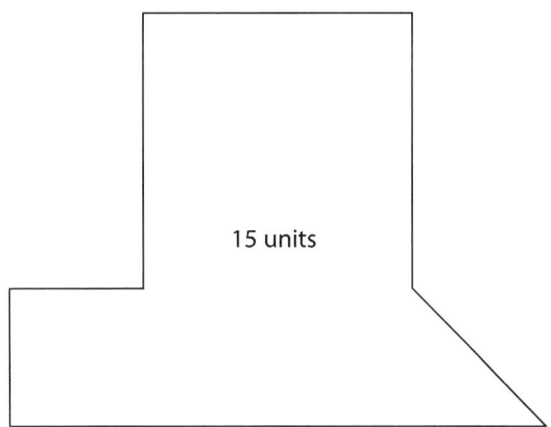

15 units

Step A: Create the shape above using pieces from a single Tangram set with a total of 15 units.

Step B: Divide it into three identical parts.

2.5.   *The Six-Pointed Star*

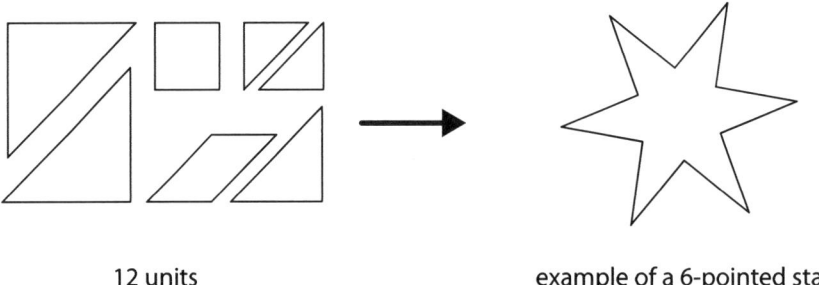

12 units   example of a 6-pointed star

Create a six-pointed star using pieces from a single Tangram set with a total of 12 units.

There are at least two solutions.

> **Note:** The shape of the six-pointed star above is just an example; the solutions need not look exactly the same.
> In this puzzle, the pieces need not align fully along their edges.

2.6.   *The Return of the Fifth Division*

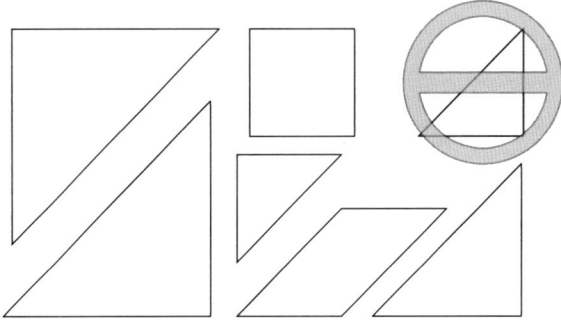

Step A: Create a symmetrical shape using the pieces from a single Tangram set with a total of 15 units.
Step B: Divide it into five identical parts.

Find at least two solutions.

2.7. *Poker*

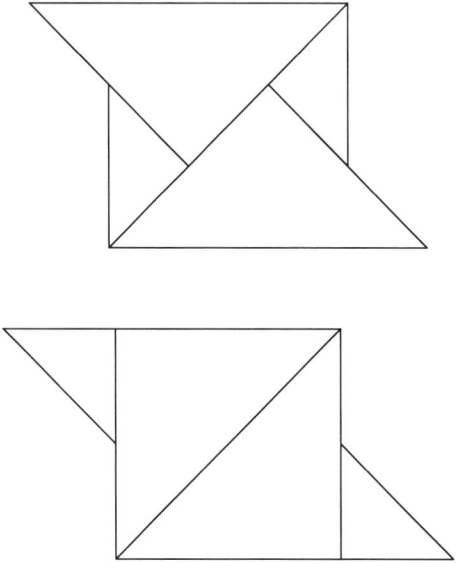

Two similar shapes are presented here. Cover each of them by using multiple copies of two-unit pieces; you may not use other pieces of different units.

Following poker rules, try to maximize your "score" by making pairs, triples, four of a kind, etc. with the units you use to cover the shapes.

Note: The ranking order in poker is: a pair; a triple; two pairs; full house; four of a kind.

2.8. *A Pair of Symmetry*

Join the parts to create a symmetrical shape.

2.9.   *The Symmetrical Balance*

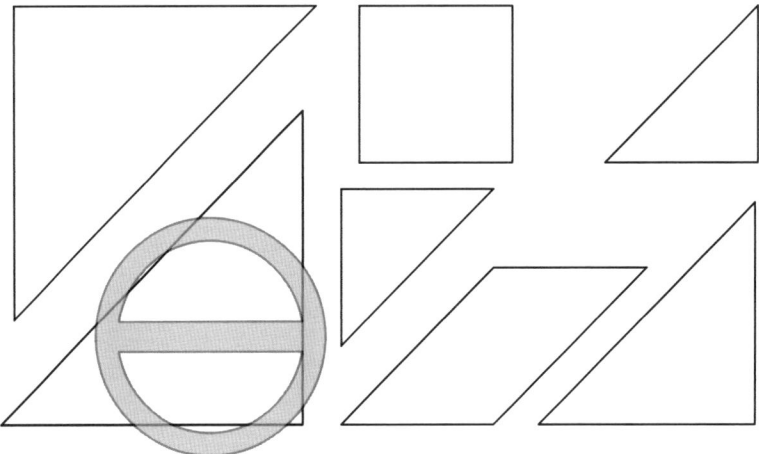

Remove one of the large triangles. Using the remaining six pieces create two identical shapes.

There are at least two solutions.

2.10.   *Twelve-Unit Rectangle*

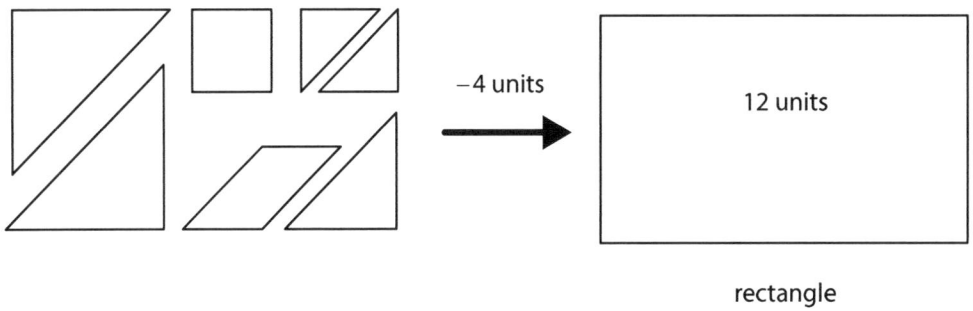

rectangle

Choose pieces with a total of 12 units and arrange them to form a rectangle.

Find at least two solutions.

2.11. *Twelve-Unit Trapezium*

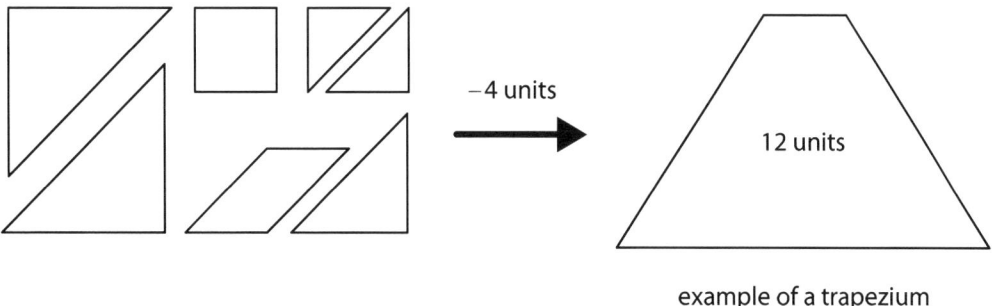

example of a trapezium

Choose pieces with a total of 12 units and arrange them to form a trapezium.

Find at least two solutions.

Note: The shape of the trapezium above is just an example; the solutions need not look exactly the same.

2.12. *The Other Pair of Symmetry*

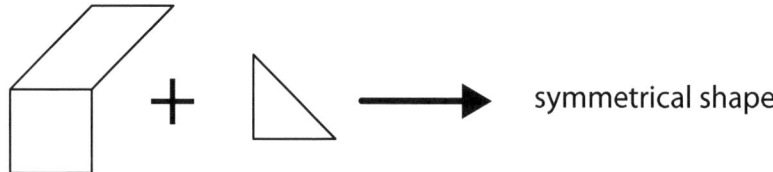

Join the two parts to create a symmetrical shape.

2.13.   *Twelve-Unit Spinning Top*

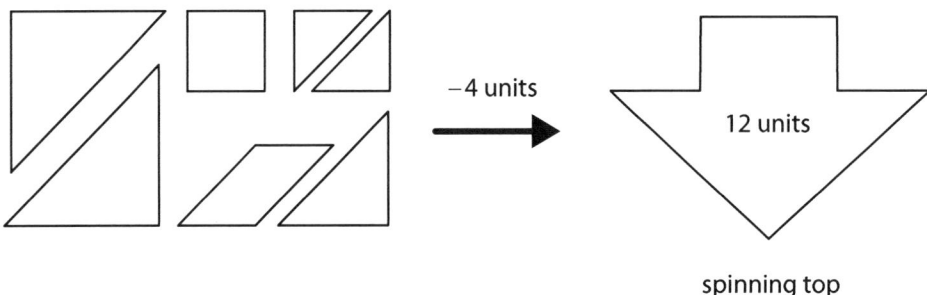

spinning top

Choose pieces with a total of 12 units and arrange them to form a spinning top.

Find at least three solutions.

Note: The shape of the spinning top above is just an example; the solutions need not look exactly the same.

2.14.   *15 by 5*

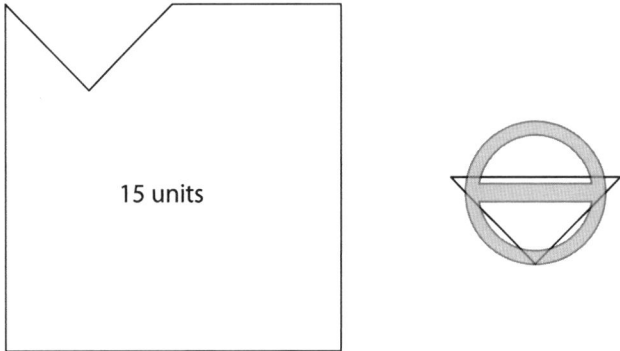

Step A: Put aside a small triangle. Arrange the remaining pieces to form the shape above.

Step B: Divide it into five identical parts.

2.15.   *Twelve-Unit Rectangle Division*

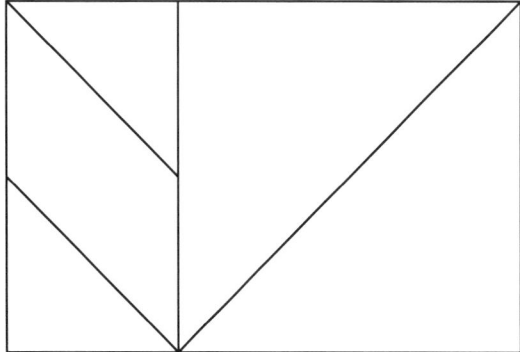

Divide the shape above into four identical parts.

2.16.   *The Hexagon Paradox*

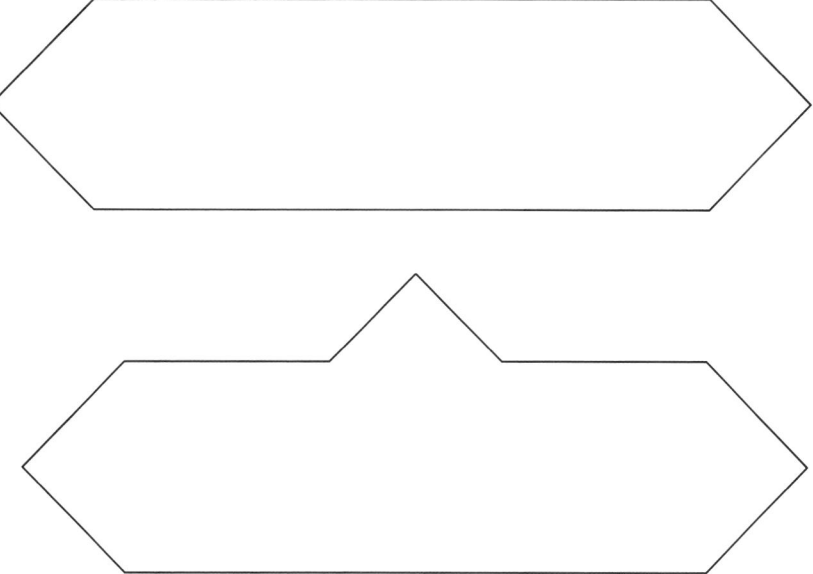

Choose six pieces to create each of the shapes above. Use the same six pieces for both shapes.

2.17. *The Small Symmetry*

By using the four different small pieces above, create a symmetrical shape.

Find at least four solutions.

2.18. *All-Triangle Division*

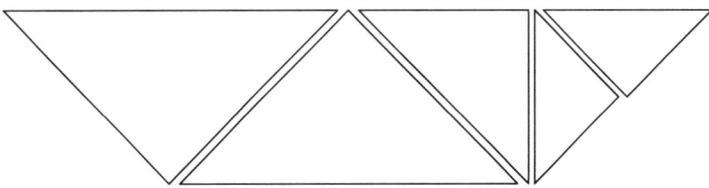

Use all the five triangle pieces. Create a shape that can be divided into four identical parts.

2.19. *Four-Piece Squares*

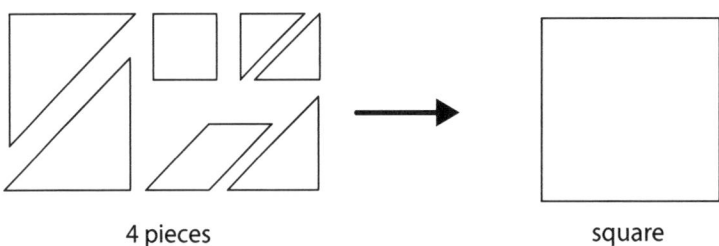

4 pieces   square

Create as many squares as you can using any four pieces.

There are three solutions.

2.20.   *Rectangles All Over*

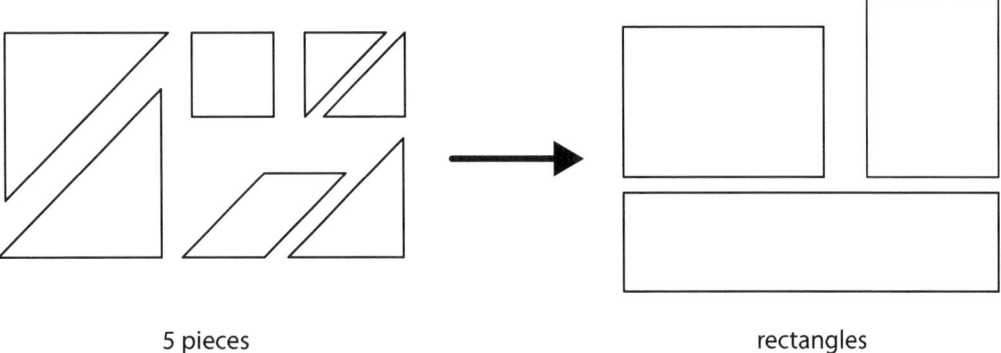

5 pieces   rectangles

Create as many rectangles as you can using any five pieces.

There are at least four solutions.

2.21.   *Symmetrical Triangles*

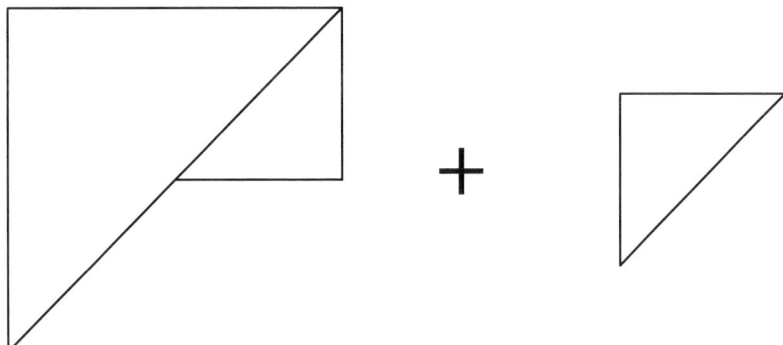

Use the pieces above to form a symmetrical shape.

Find at least four solutions. Describe how each solution is unique.

2.22. *The Maximum Four*

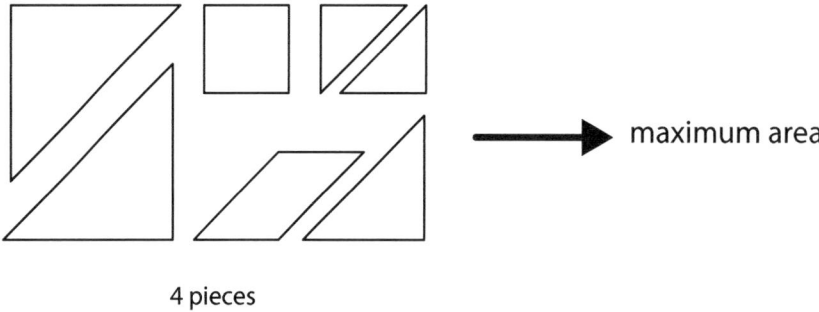

maximum area

4 pieces

Using any four pieces, create either a square, a triangle or a parallelogram, with the largest area possible.

2.23. *The Minimum Five*

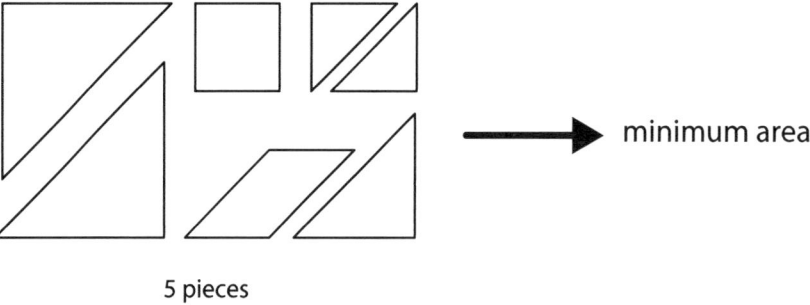

minimum area

5 pieces

Using any five pieces, create either a square, a triangle or a parallelogram, with the smallest area possible.

2.24.　*Pentominoes!*

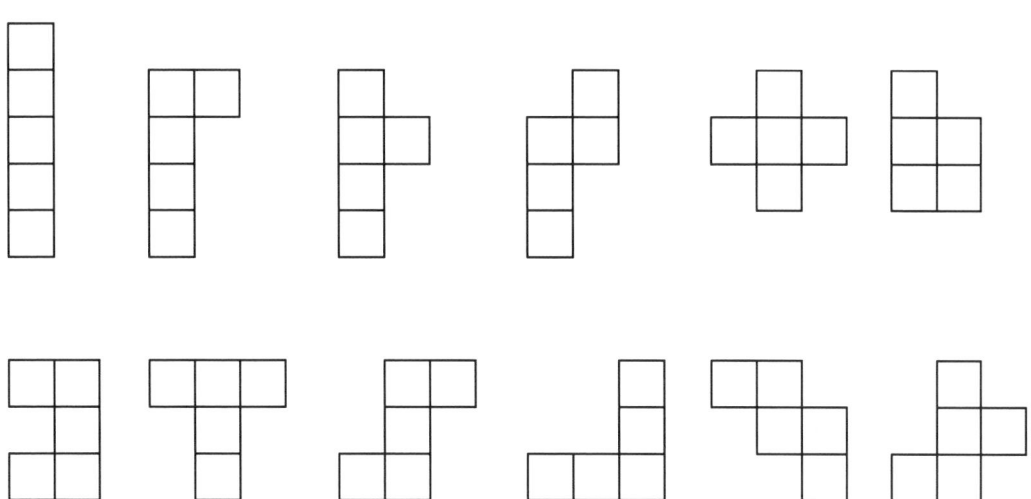

Pentominoes are shapes that are created with five squares connected edge-to-edge. There are altogether 12 pentominoes.

Here, every square of the pentomino is made up of two units, therefore 10 units are needed.

Try to recreate each of the pentominoes shown above by using a partial Tangram set.

Some pentominoes do not have a solution!

2.25. *Cube Nets*

A cube net is a two-dimensional shape that can be folded to form a three-dimensional cube. Use any six pieces of a Tangram set to recreate the cube nets shown above.

Here, every square of the nets is made up of two units, therefore 12 units are needed.

Note: For some, there is no perfect solution, but you can still create a cube net that is off by only one triangle.

2.26. *Six-Unit Division*

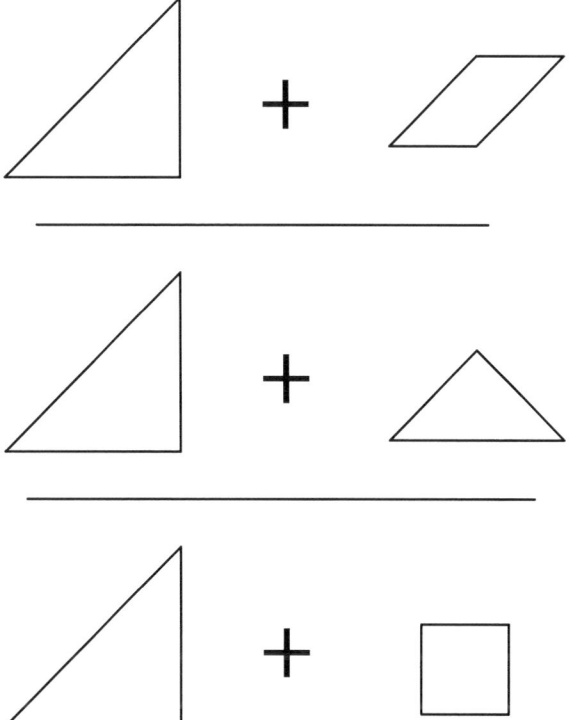

Combine the large triangle with a single two-unit piece to form three different shapes, each of which can be divided into two identical parts.

2.27. *Four-Unit Combos*

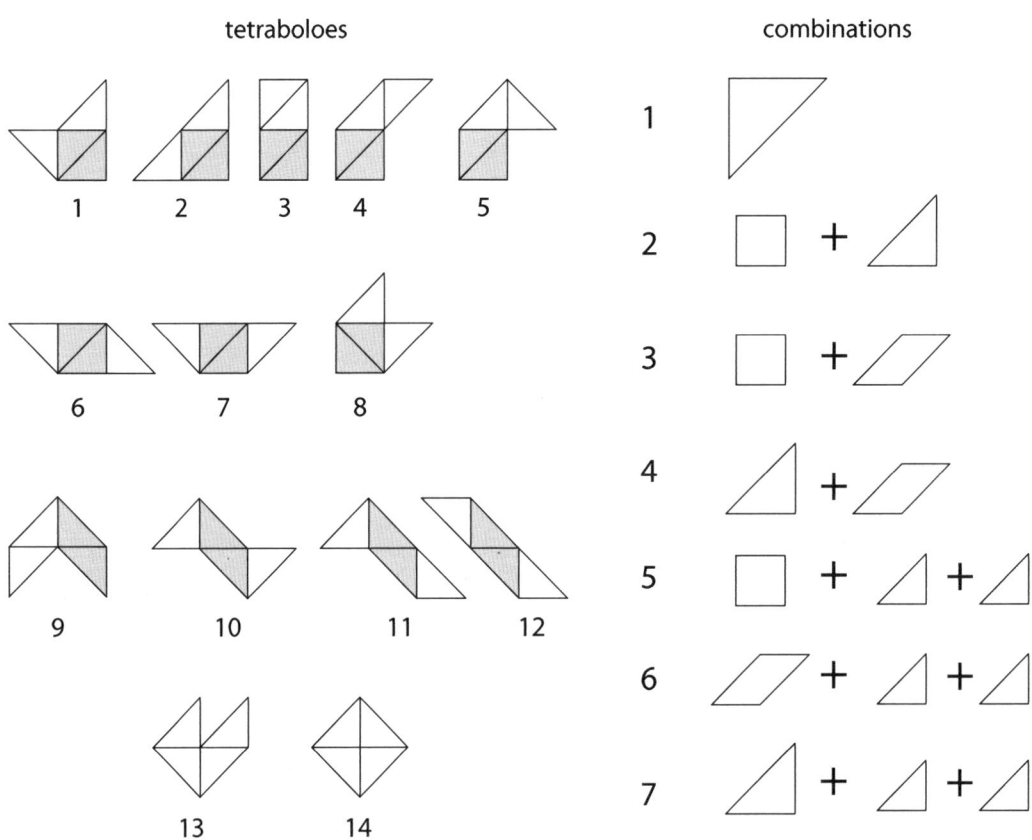

Tetraboloes are shapes that are created with four right-angled isosceles triangles. There are altogether 14 tetraboloes.

There are seven possible combinations of Tangram pieces that will cover the tetraboloes. Find which one of the seven combinations will cover the most tetraboloes.

2.28. *Five-Unit Combos*

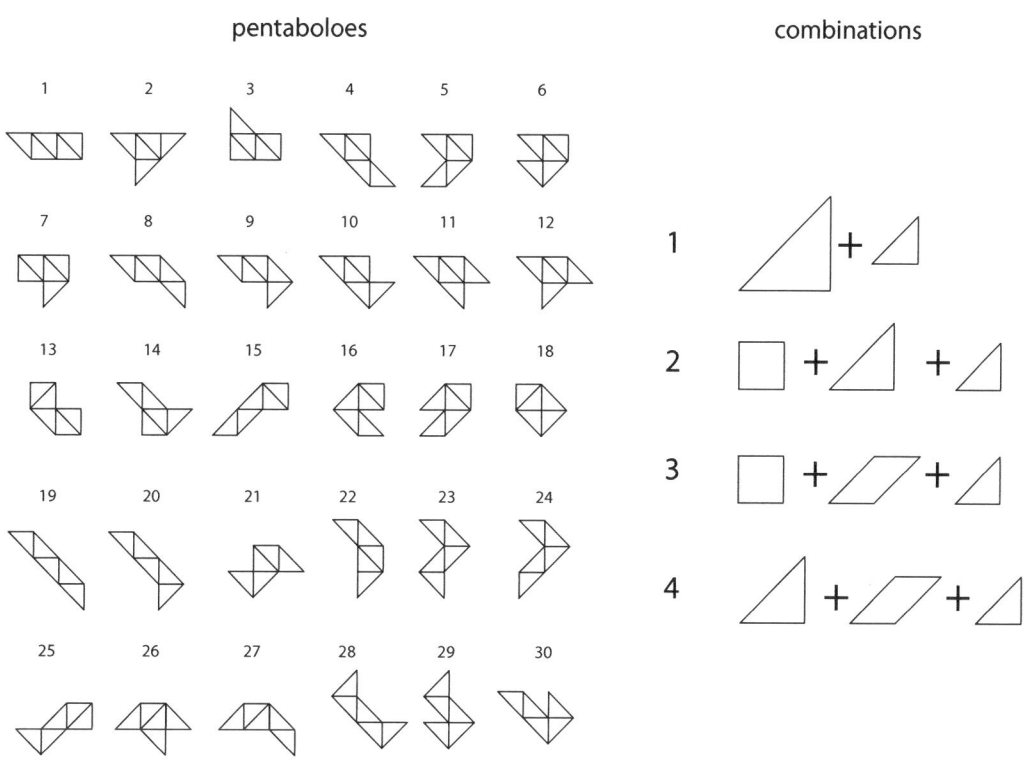

pentaboloes

combinations

Pentaboloes are shapes that are created with five right-angled isosceles triangles. There are altogether 30 pentaboloes.

There are four possible combinations of Tangram pieces that will cover the pentaboloes. Find which one of the four combinations will cover the most pentaboloes.

2.29. *The Missing Square*

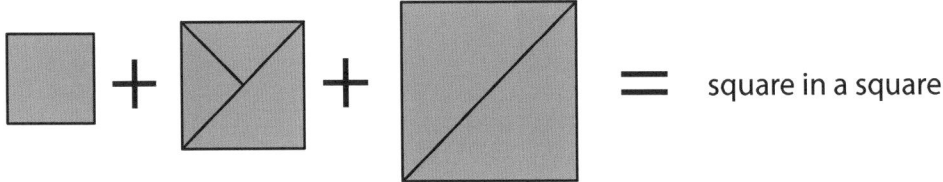

With three squares, made up of six pieces, create a square within a square.

2.30. *The Negative Puzzle*

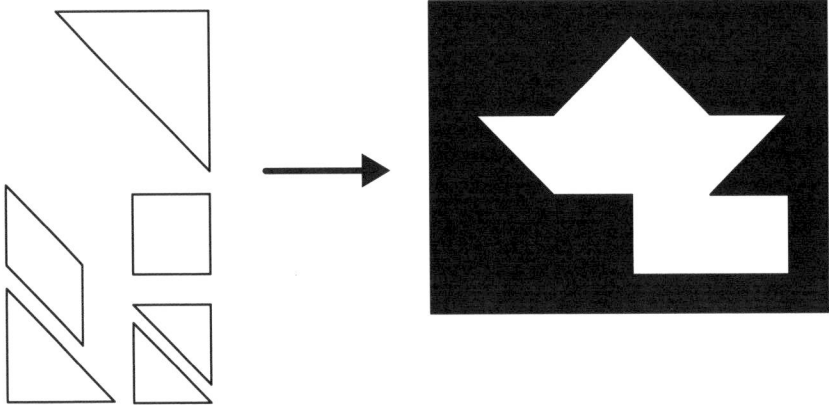

Locate the position of just one of the six pieces in the white space so that the positions of the remaining pieces are revealed.

2.31. *The Magnificent Seven*

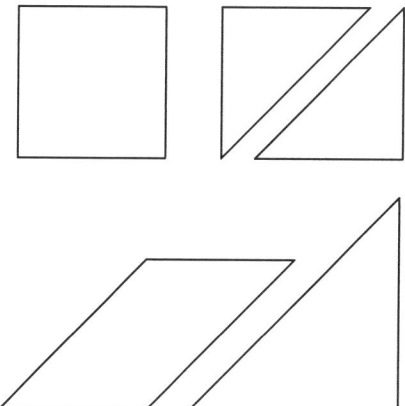

Using only the small pieces of the set, create seven possible convex shapes.

2.32. *Symmetry All Over*

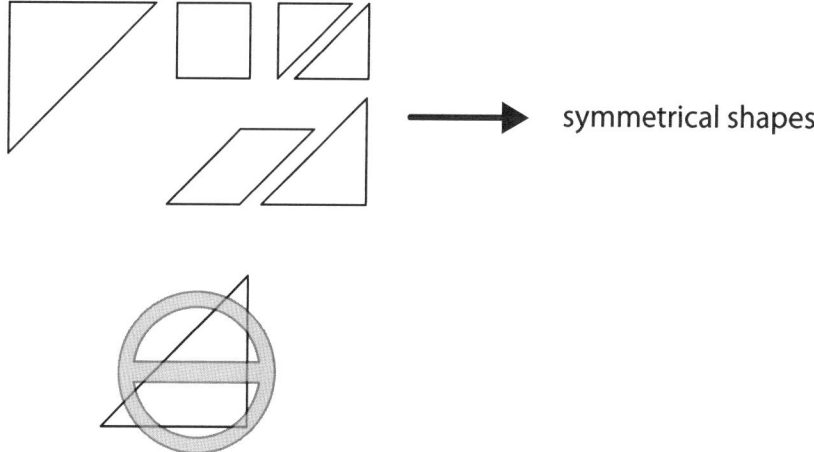

Using the six pieces above, create symmetrical shapes.

Bonus points for solutions that have more than one symmetry axis.

There are at least two solutions.

2.33.   *Spin or Reflect*

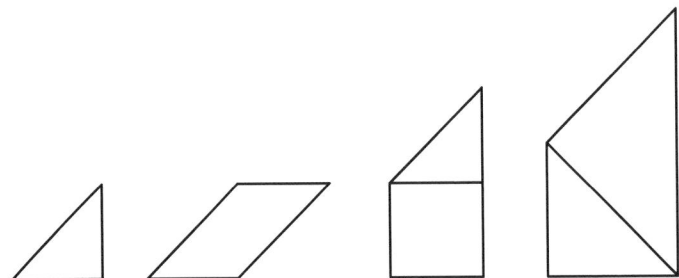

Using two or more of the pieces and parts above, create shapes that have rotational or reflective symmetry.

Find at least one solution that consists of all the four pieces/parts.

Solutions

2.1. *Six-Piece Rectangle*

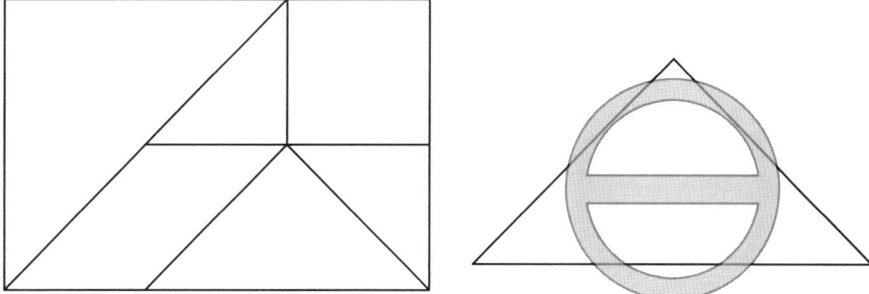

2.2. *The Fifth Division — Diamond*

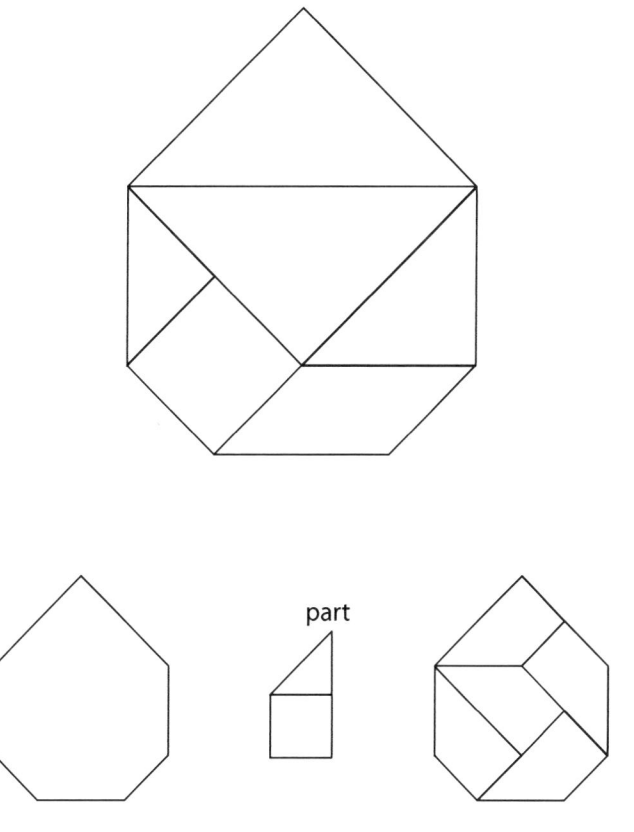

part

2.3.   *The Butterfly*

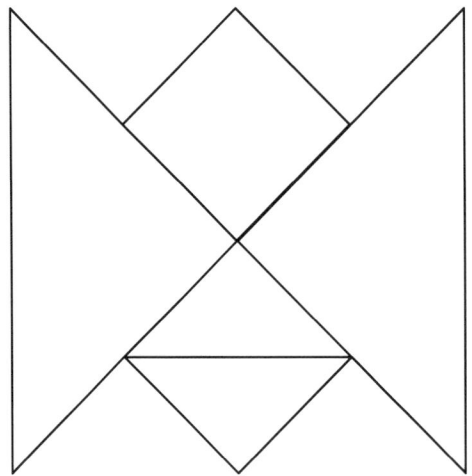

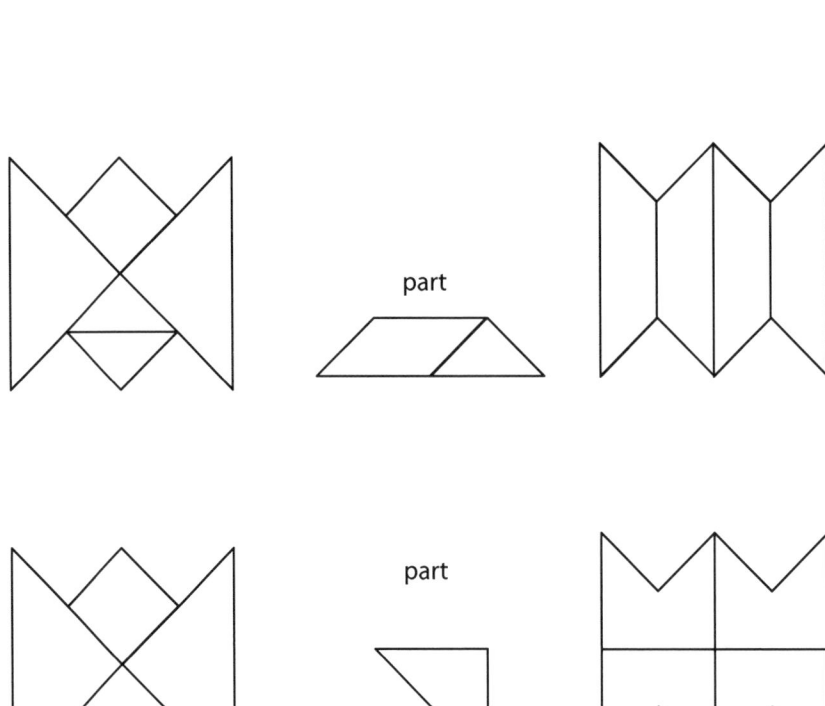

2.4. *The Third Division*

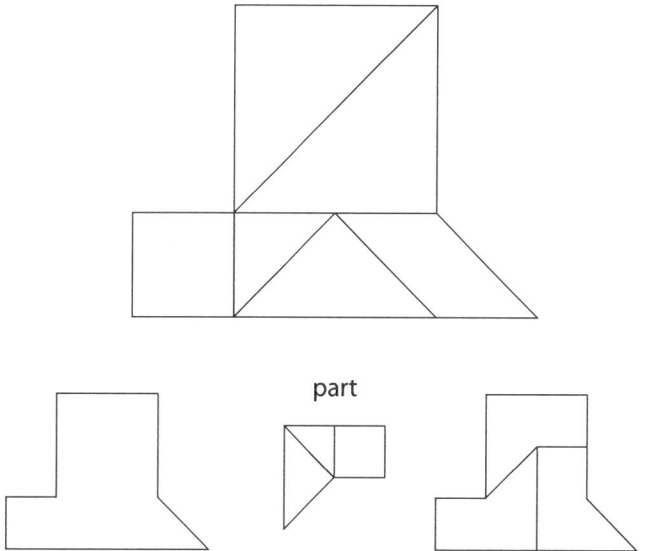

part

2.5. *The Six-Pointed Star*

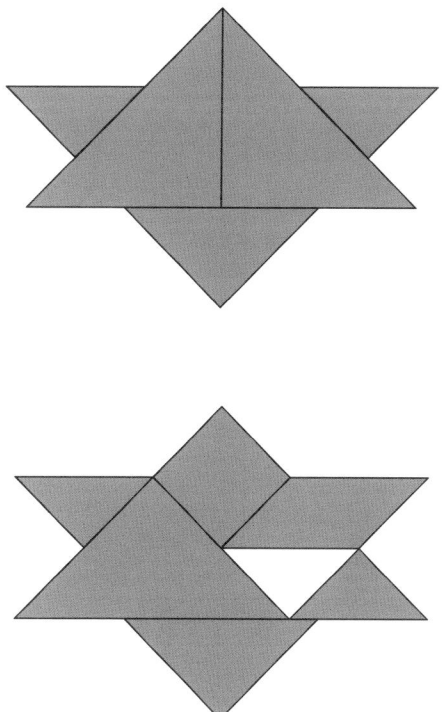

2.6. *The Return of the Fifth Division*

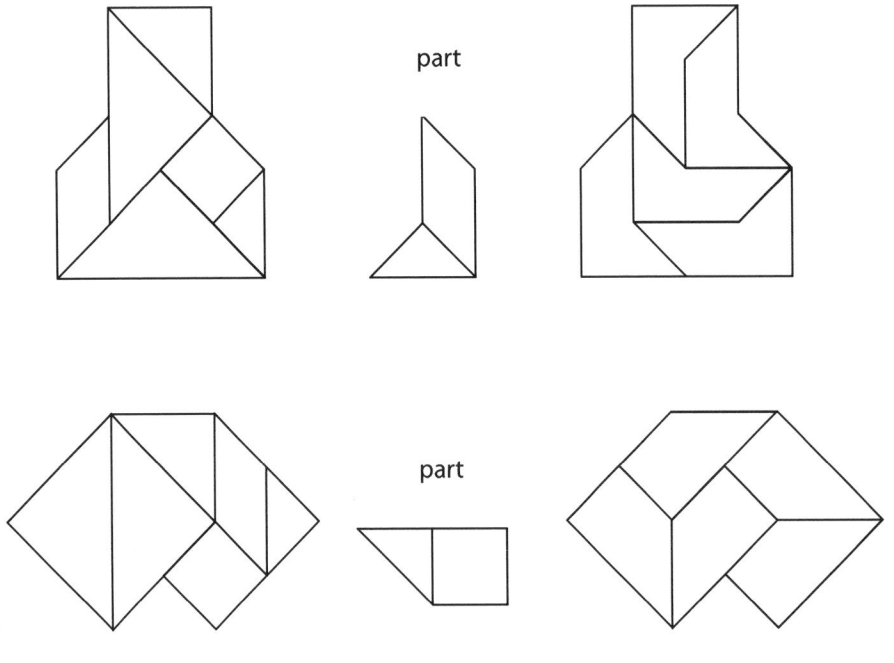

2.7. *Poker*

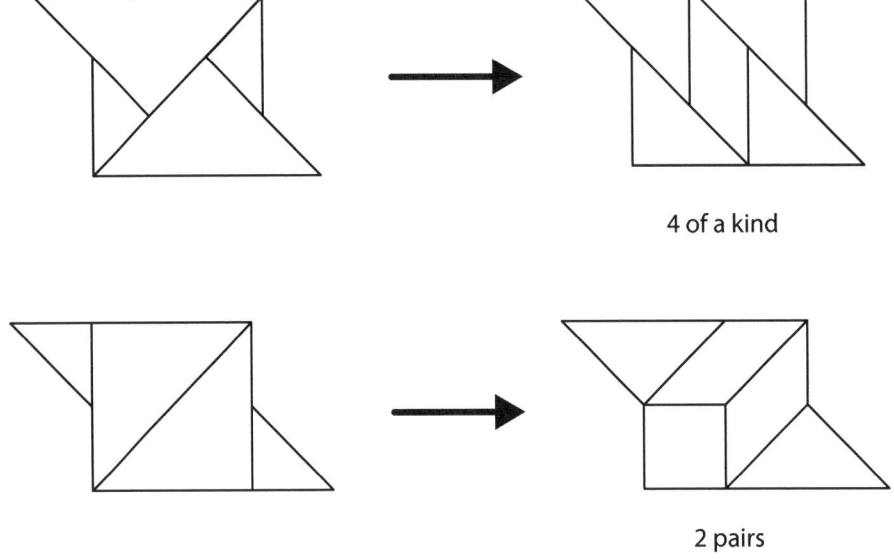

4 of a kind

2 pairs

2.8. *A Pair of Symmetry*

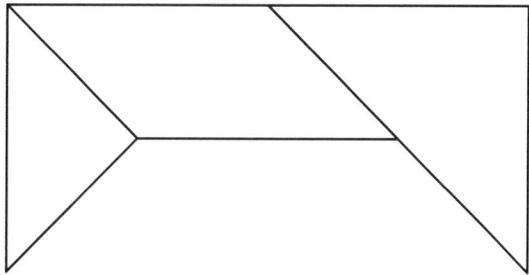

2.9. *The Symmetrical Balance*

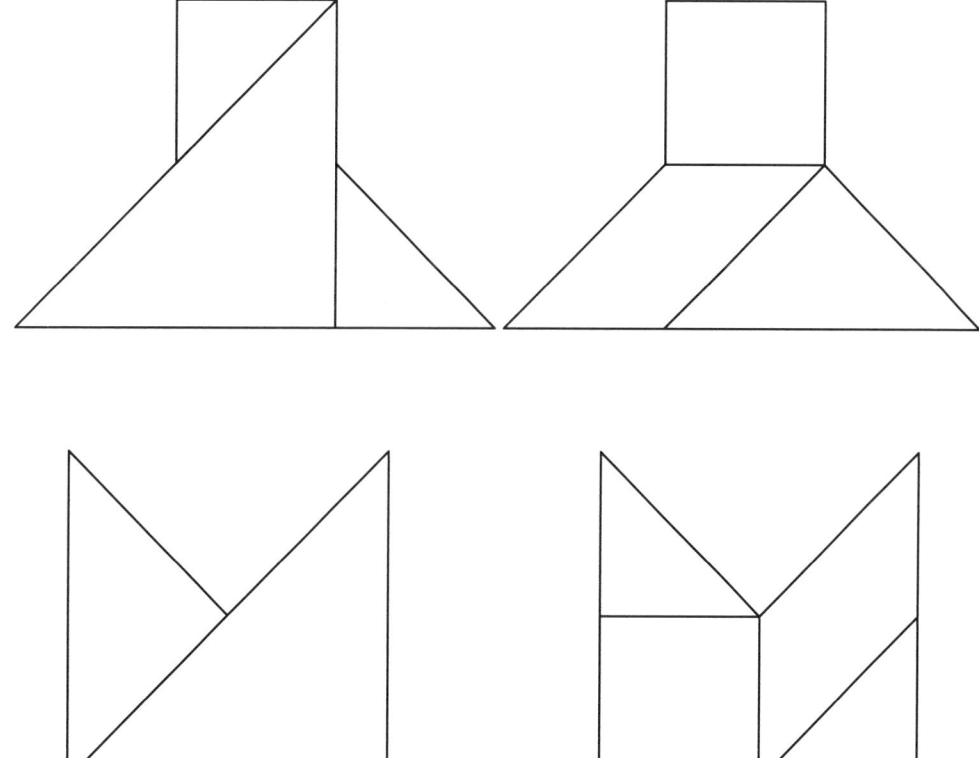

2.10.   *Twelve-Unit Rectangle*

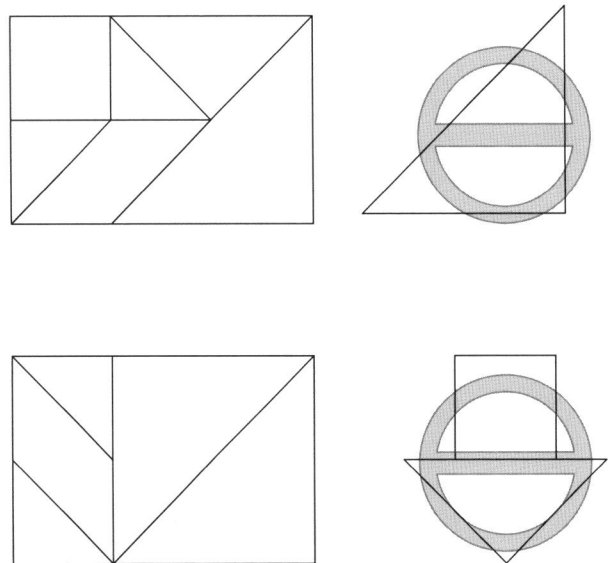

2.11.   *Twelve-Unit Trapezium*

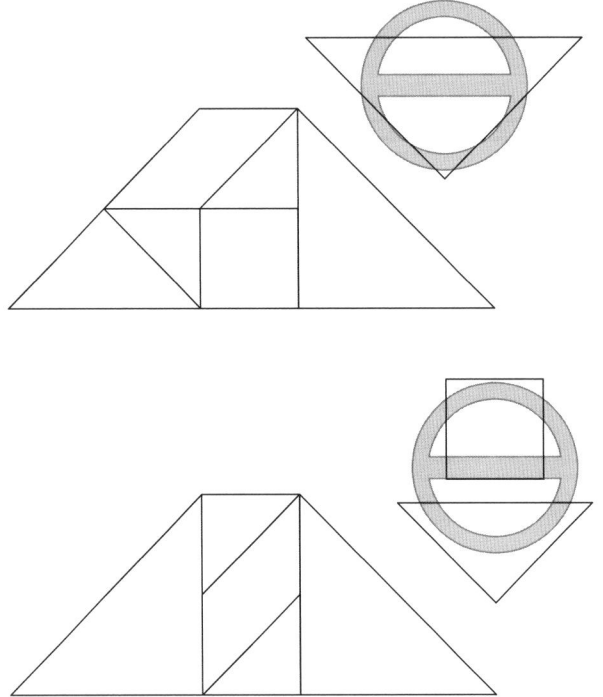

2.12. *The Other Pair of Symmetry*

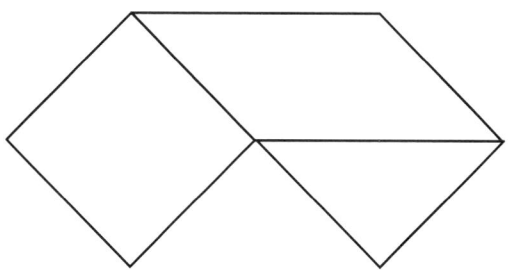

2.13. *Twelve-Unit Spinning Top*

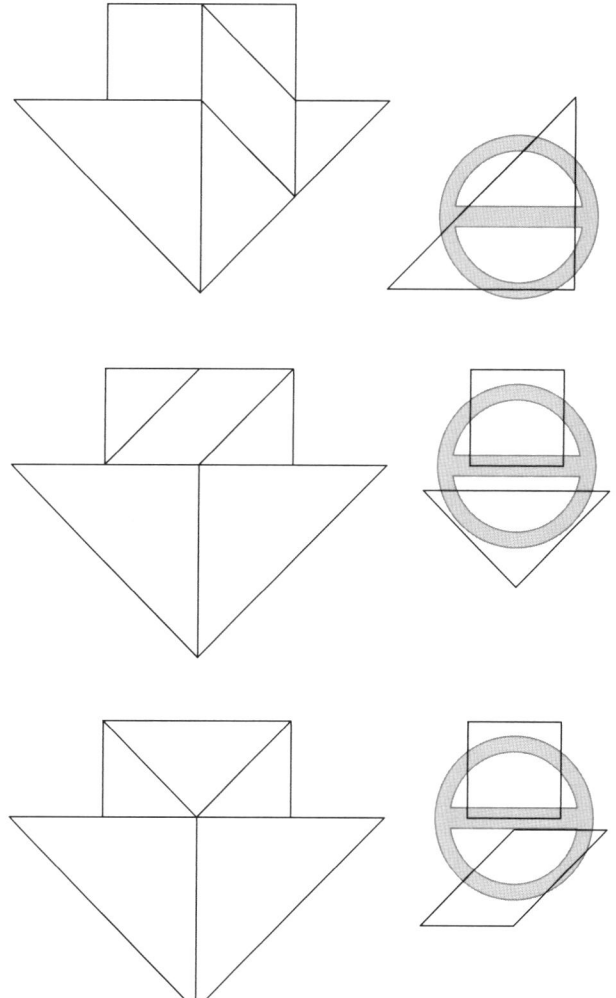

2.14.   *15 by 5*

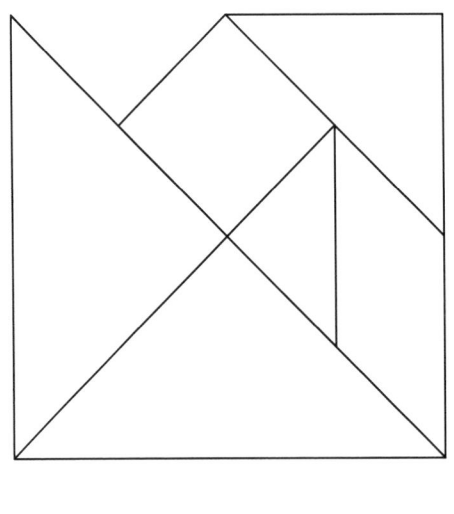

parts

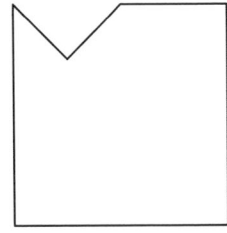 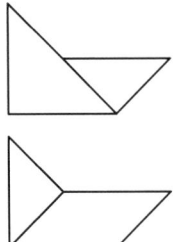 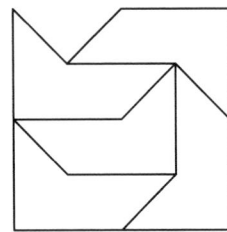

2.15.   *Twelve-Unit Rectangle Division*

parts

 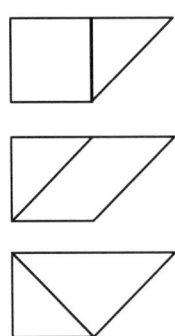 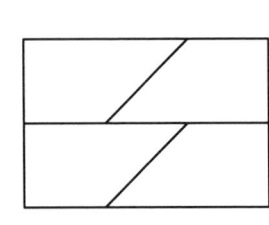

2.16.   *The Hexagon Paradox*

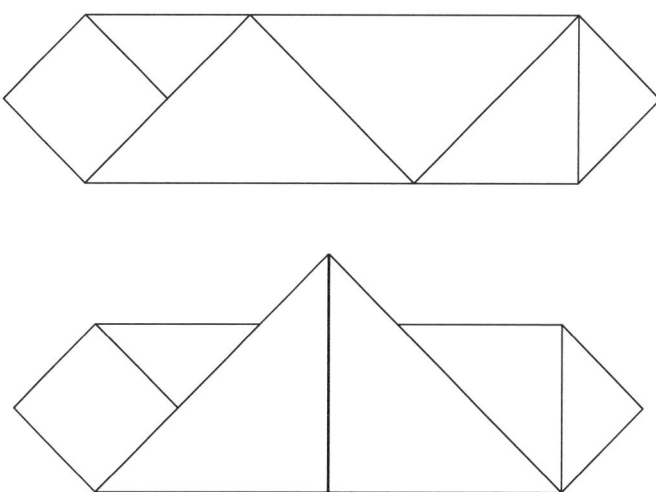

2.17.   *The Small Symmetry*

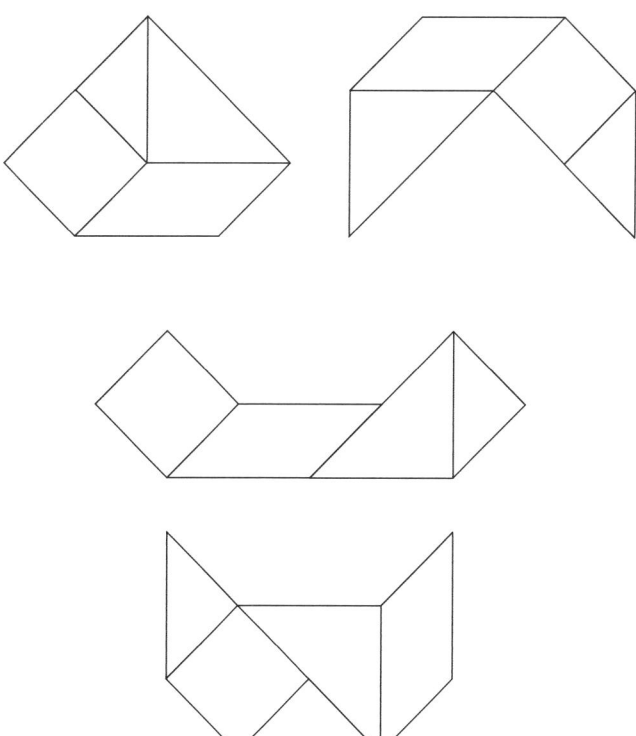

2.18.   *All-Triangle Division*

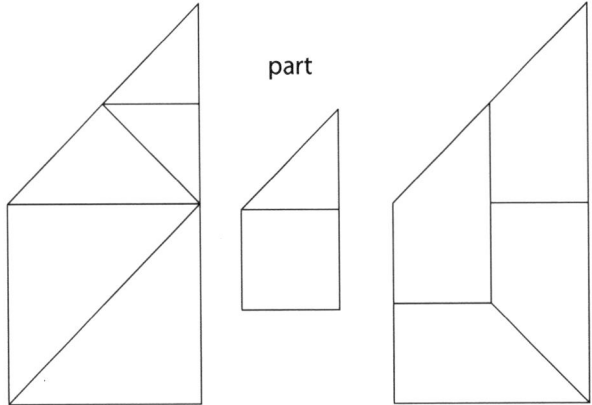

part

2.19.   *Four-Piece Squares*

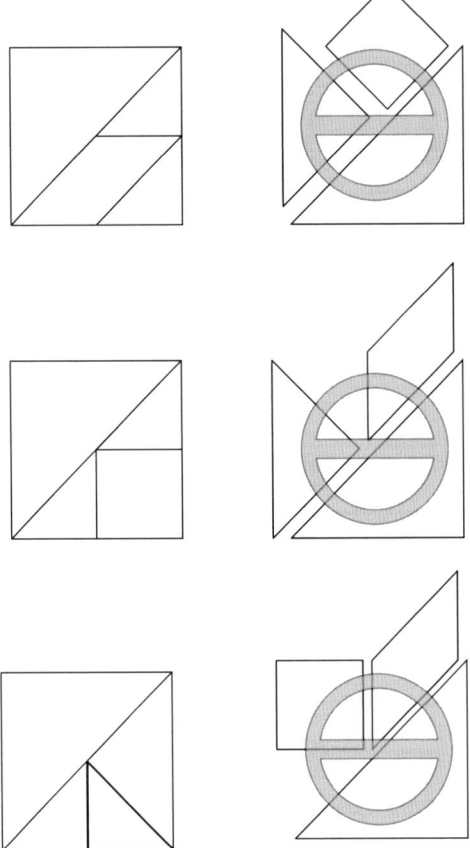

2.20.   *Rectangles All Over*

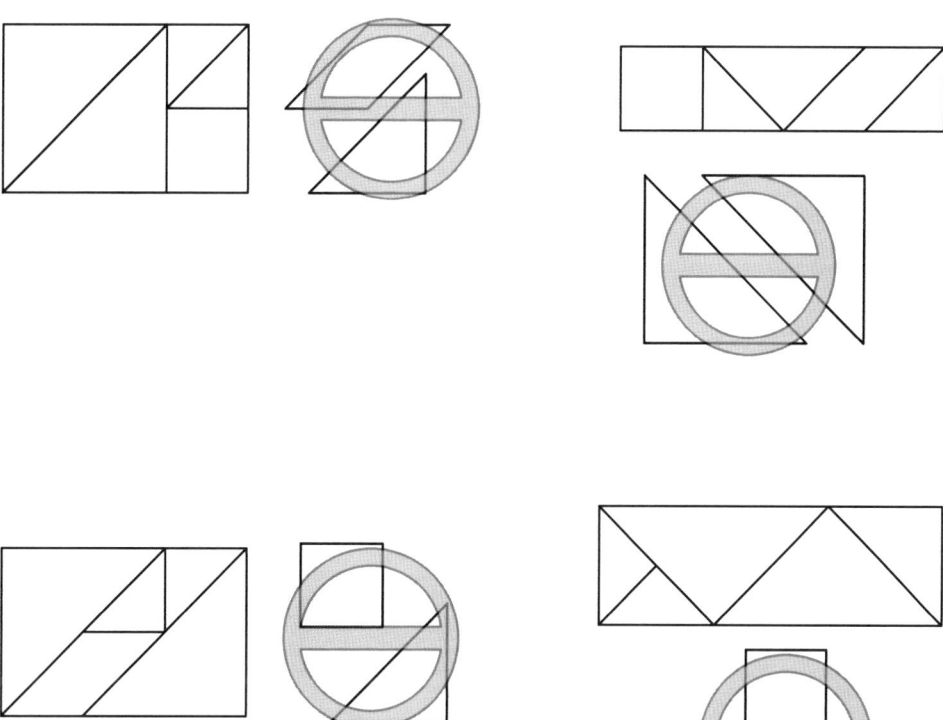

2.21.   *Symmetrical Triangles*

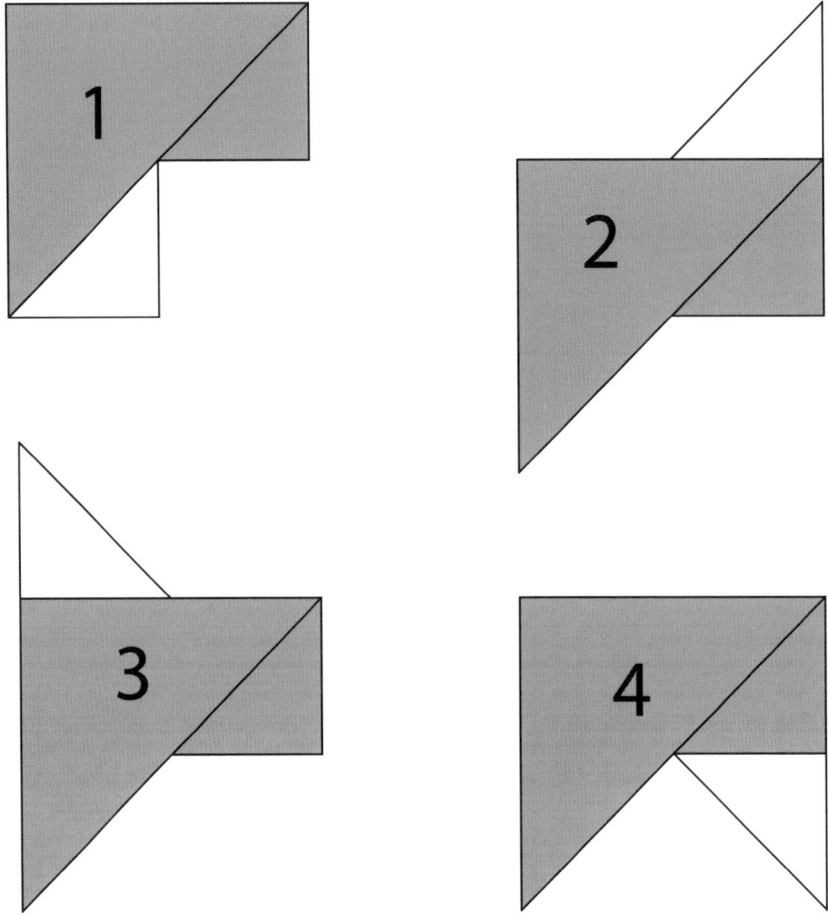

Shape 1:   The shortest perimeter (short edges totalling eight units).

Shape 2:   The only one which has a rotational symmetry.

Shape 3:   The only one which has an edge that is the length of three units.

Shape 4:   The only one which has five corners (all the other shapes have six corners). The sum of its inner angles is 540° (for the other shapes, the sum of the inner angles is 720°).

2.22. *The Maximum Four*

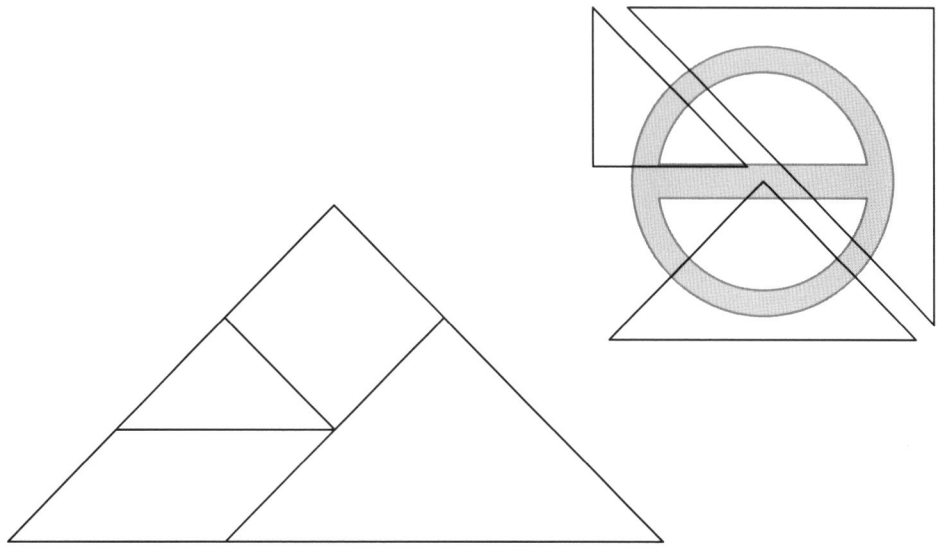

The maximum area is nine units!

2.23. *The Minimum Five*

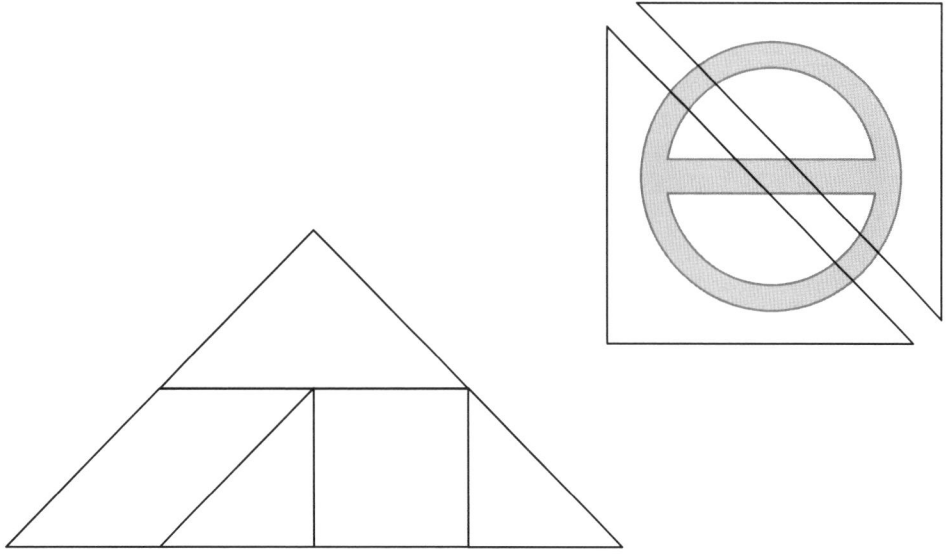

The minimum area is eight units!

2.24.   *Pentominoes!*

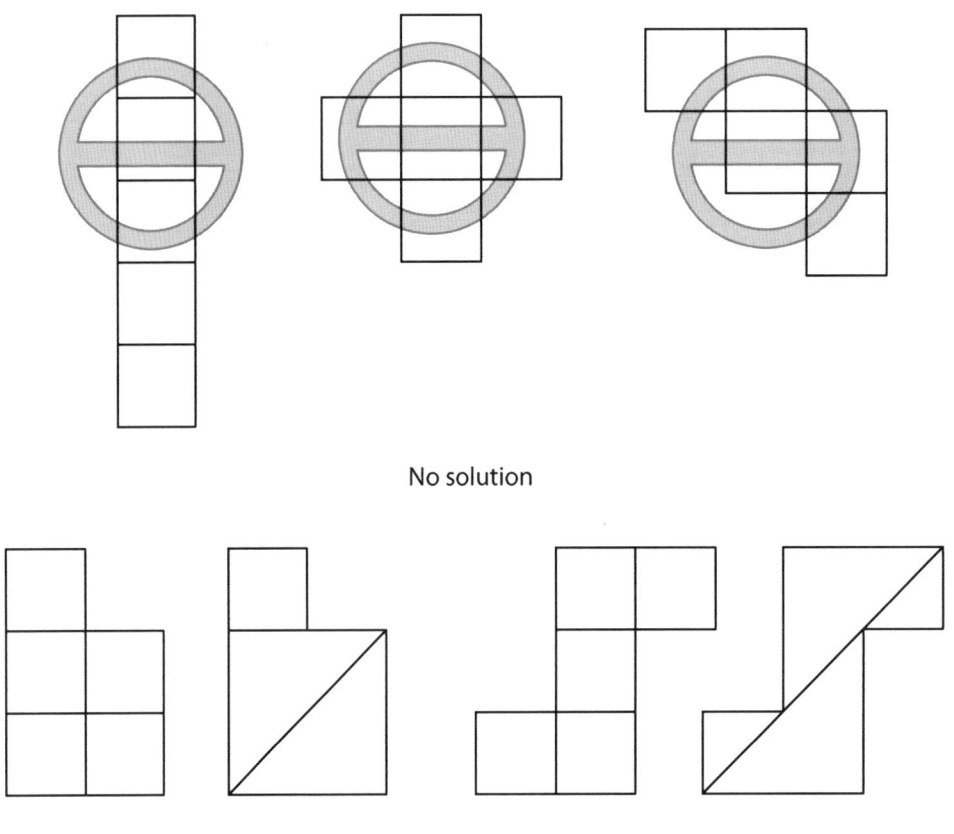

No solution

3 pieces   4 pieces

5 pieces

2.25.   *Cube Nets*

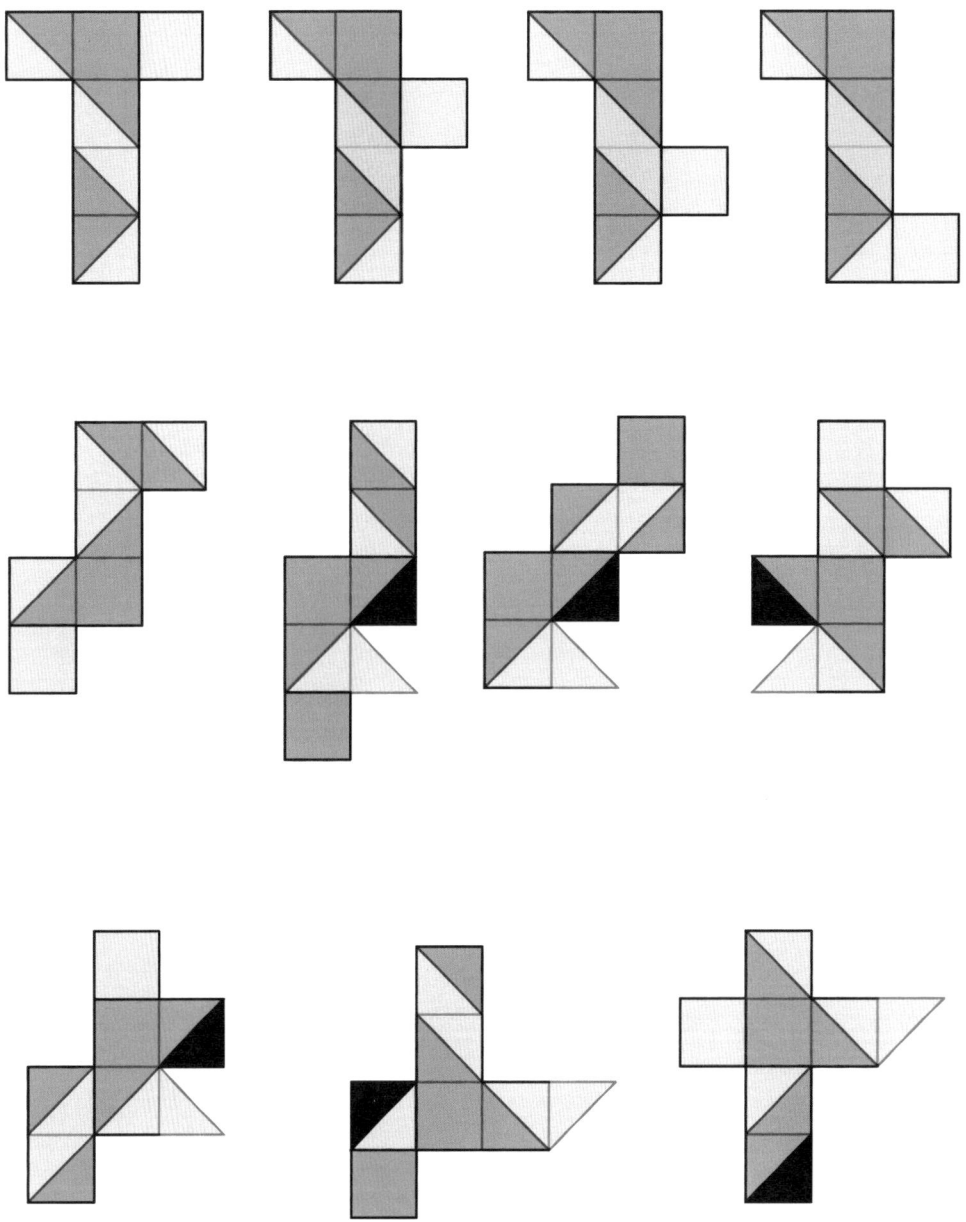

The black triangle is part of the original cube net. Although it cannot be covered by the Tangram pieces, the one extra triangle can cover it, when the cube net is folded into a cube.

2.26. *Six-Unit Division*

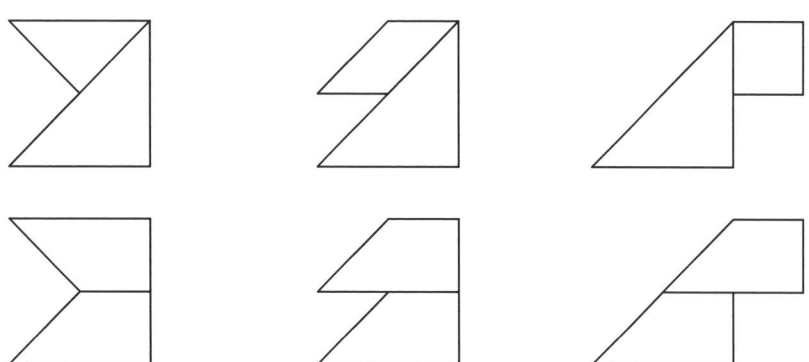

2.27. *Four-Unit Combos*

1 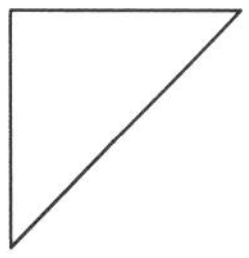   Shape 2 (one shape only)

2 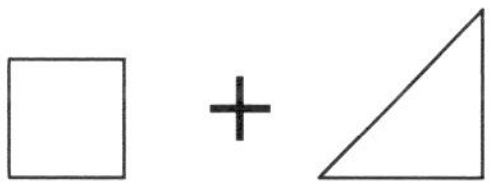   Shape 5 (one shape only)

3 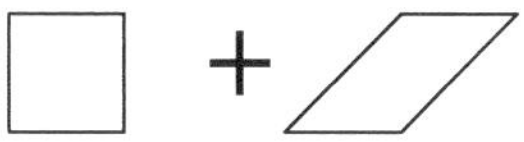   Shape 4 (one shape only)

4 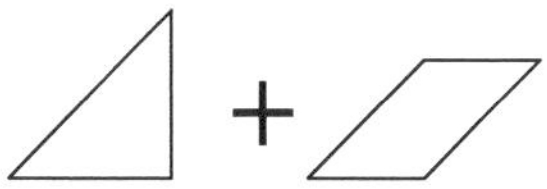   Shapes 11, 13 (two shapes)

5 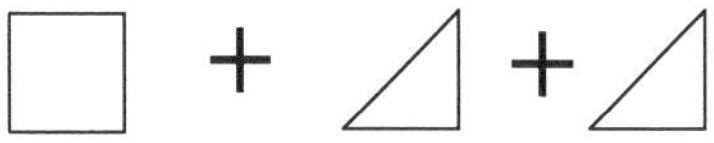   Shapes 1, 2, 3, 4, 5, 6, 7, 8
     (eight shapes)

6    Shapes 1, 2, 3, 4, 5, 6, 7, 8, 9, 10, 11,
     12, 13 (13 shapes)

7 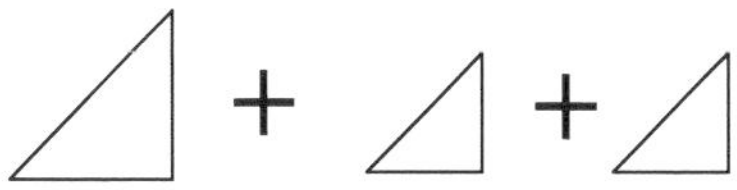   Shapes 1, 2, 3, 4, 5, 6, 7, 8, 9, 10, 11,
     13, 14 (13 shapes)

2.28.   *Five-Unit Combos*

1   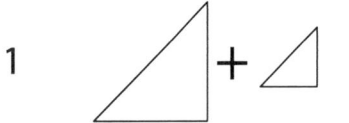   Shapes 2, 3, 4, 10, 11 (five shapes)

2   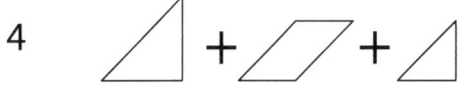   Shapes 1, 3, 6, 7, 9, 10, 13, 16, 18, 21,
                             25 (11 shapes)

3   Shapes 1, 3, 4, 5, 7, 8, 13, 14, 15, 16,
                            17, 22, 25, 27 (14 shapes)

4   Shapes 1, 2, 4, 5, 6, 7, 8, 9, 10, 12, 14,
                            15, 16, 17, 18, 20, 21, 22, 23, 24, 25,
                            26, 27, 28, 29, 30 (26 shapes)

Shape 19 does not have any solution!

2.29.   *The Missing Square*

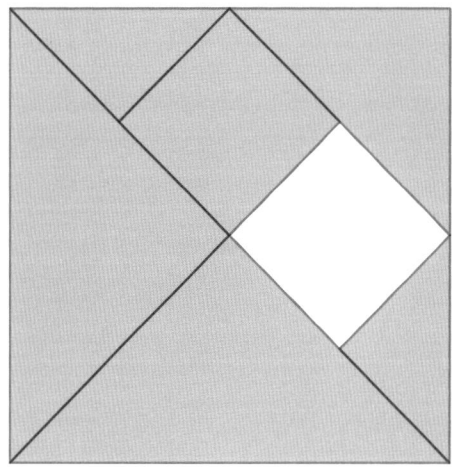

2.30. *The Negative Puzzle*

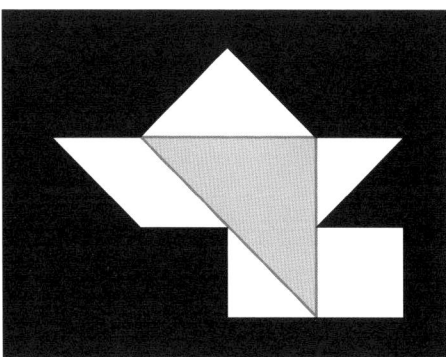

2.31. *The Magnificent Seven*

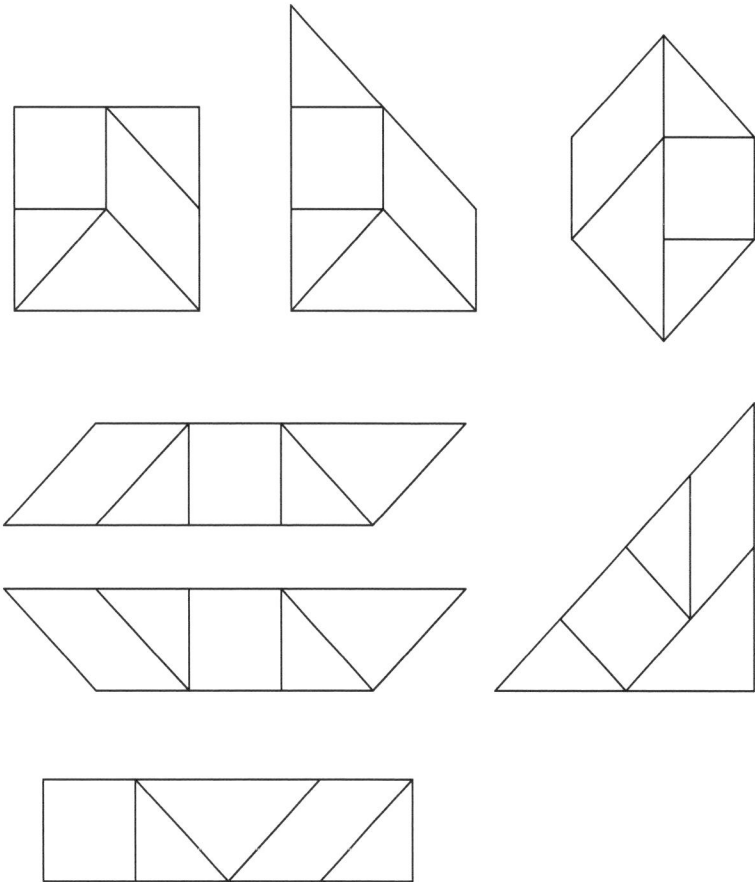

2.32. *Symmetry All Over*

 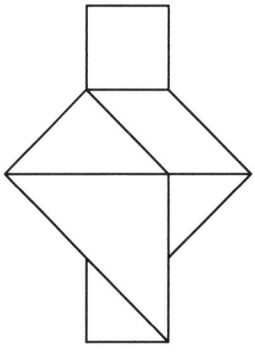

Two symmetry axes.

2.33. *Spin or Reflect*

Shapes with rotational symmetry

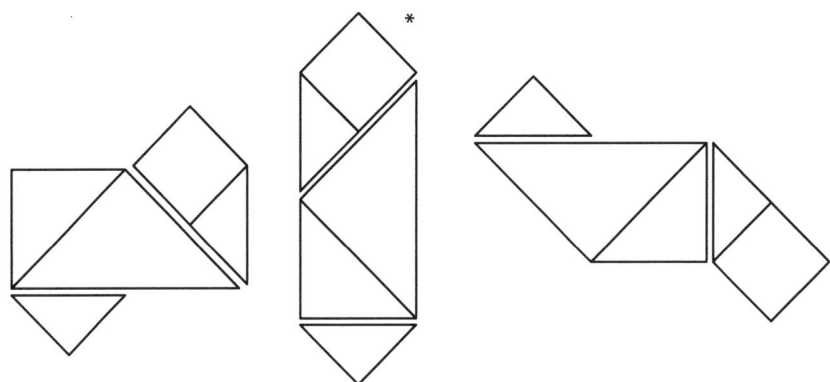

*Has both rotational and reflective symmetry.

Shapes with reflective symmetry

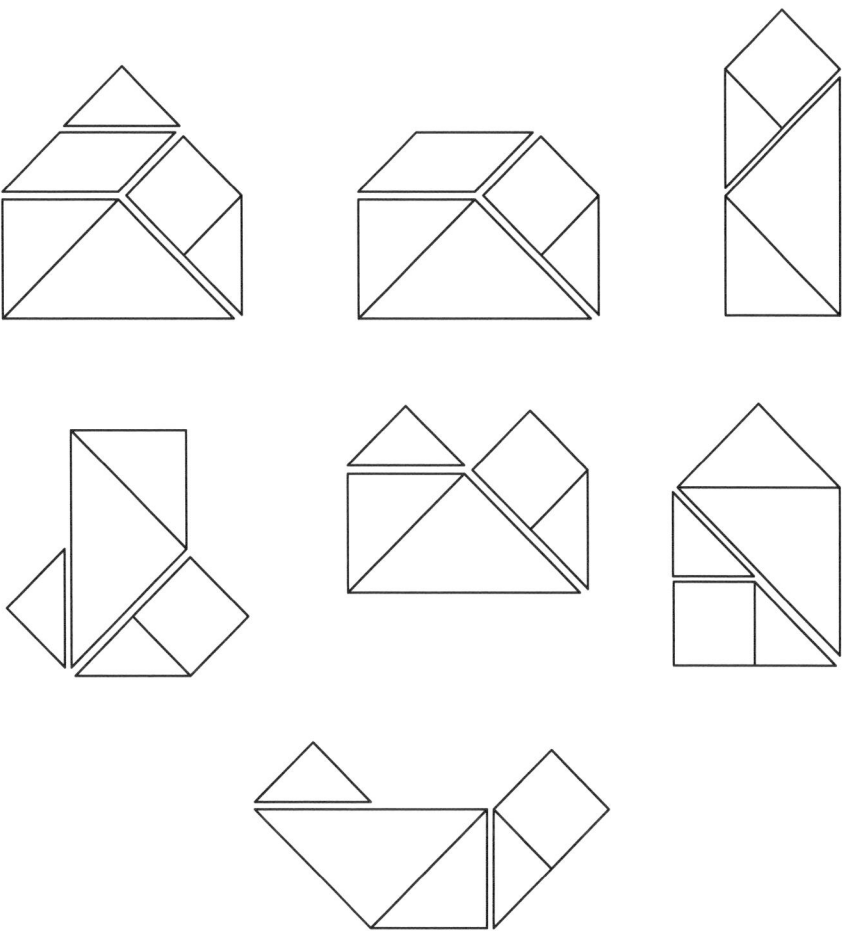

Chapter 3

Multiple-Set Tangram Puzzles

In this chapter, some puzzles need more than one set of Tangram pieces; some may need up to three sets.

Each set is presented in a different shade: white, light gray and dark gray.

3.1. *The Square Deal*

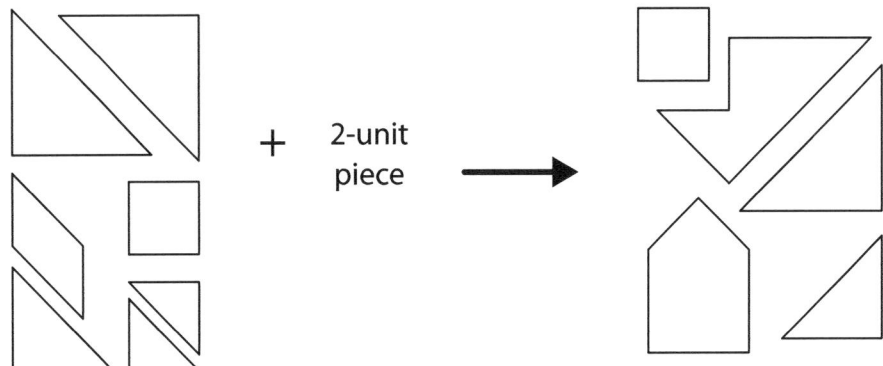

Use a full set, and add to it a two-unit piece (18 units in total).

Step A: Create the five parts shown above.
Step B: Using these five parts, create all the shapes below.

3.2.   *XL Six-Pointed Star*

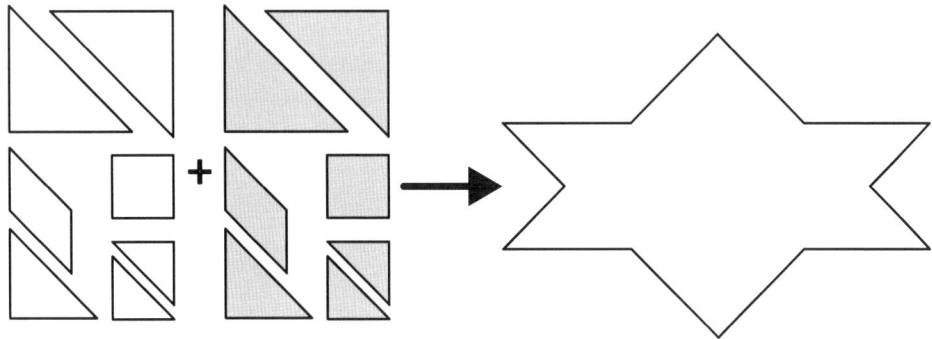

Use two sets to create this XL six-pointed star.

3.3.   *The XL Square*

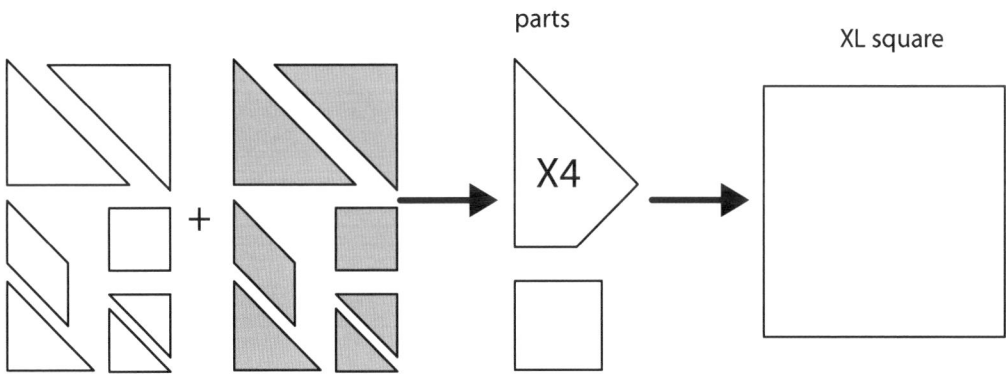

Step A: Using two sets, create five parts, of which four are identical.
Step B: With the five parts, create the XL square.

3.4.   *Max Spinners*

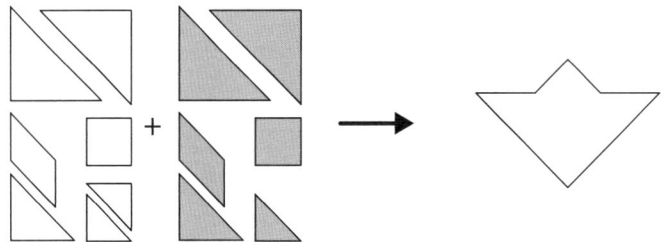

With the pieces given above, create as many spinners as you can. Each piece can be used only once.

Note: In this puzzle, the pieces need not align fully along their edges.

3.5.   *The Pentomino Rectangle*

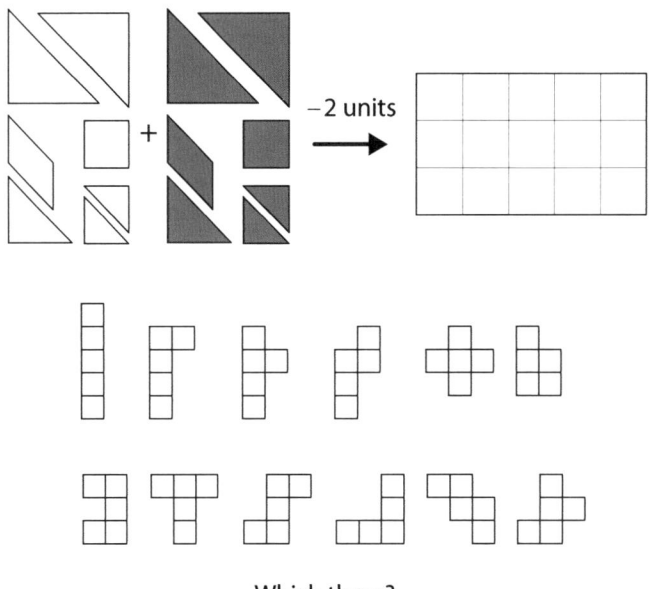

Which three?

Create a 3×5 rectangle, with two sets, minus a two-unit piece. You must be able to divide this rectangle into three valid pentominoes. The outline of the pentomino must be aligned with the outline of the pieces.

3.6. *The Largest Pentomino Rectangle*

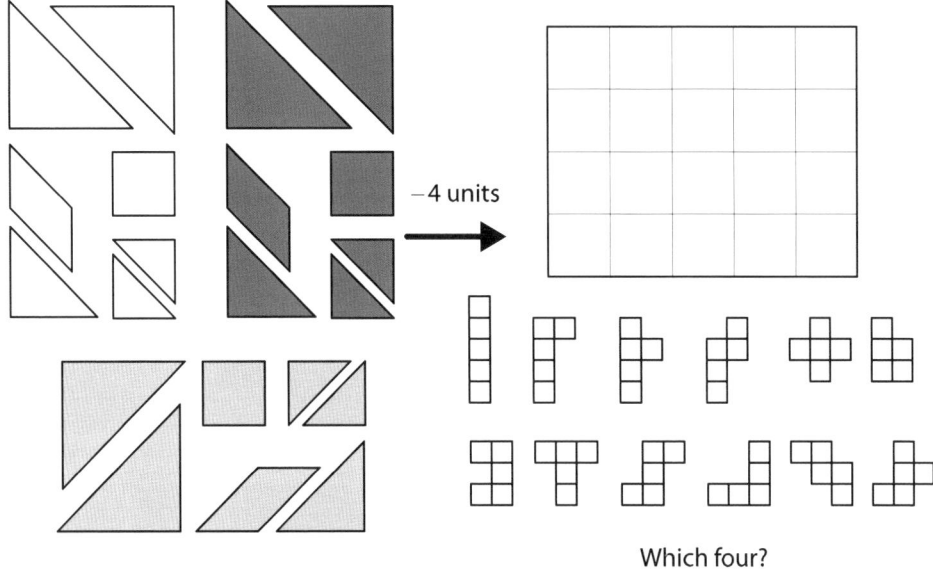

Which four?

Create a 4 × 5 rectangle, with three sets, minus two four-unit pieces. You must be able to divide this rectangle into four valid pentominoes. The outline of the pentomino must be aligned with the outline of the pieces.

3.7. *The Parrot*

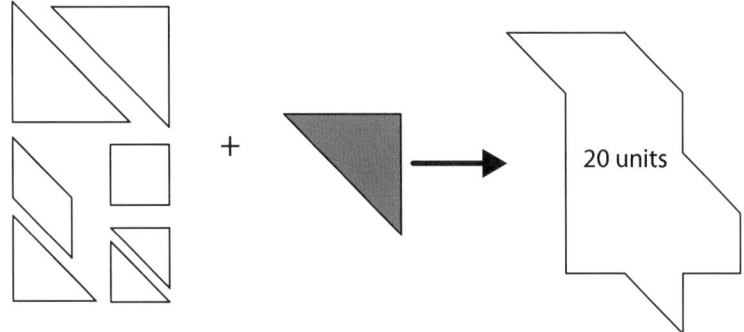

Step A: Create the shape of the parrot, with a set and an additional four-unit piece.

Step B: Divide this shape into five identical parts.

3.8.   *Square and Octagon*

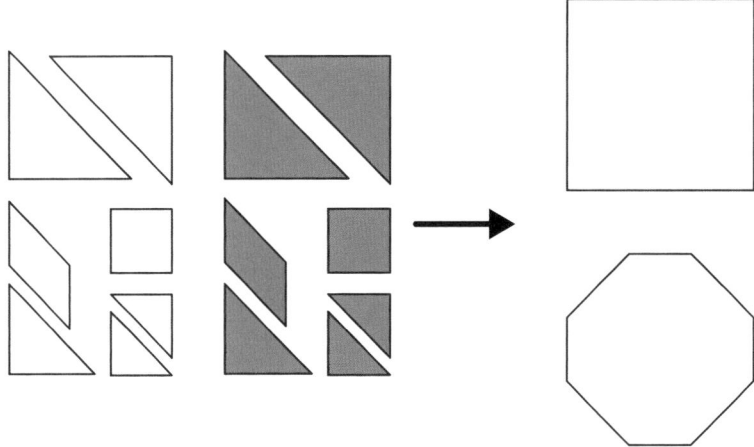

Create a square and an octagon using two sets.

3.9.   *Five, Six, Seven*

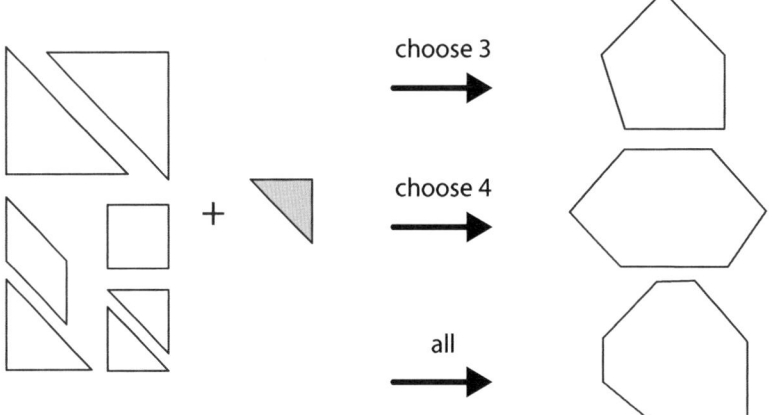

With a set and a small triangle, choose three pieces to create a pentagon; four pieces to create a hexagon, and all pieces to create a heptagon.

Note: The shapes above are just examples. There are several solutions.

3.10. *Seven to Eight*

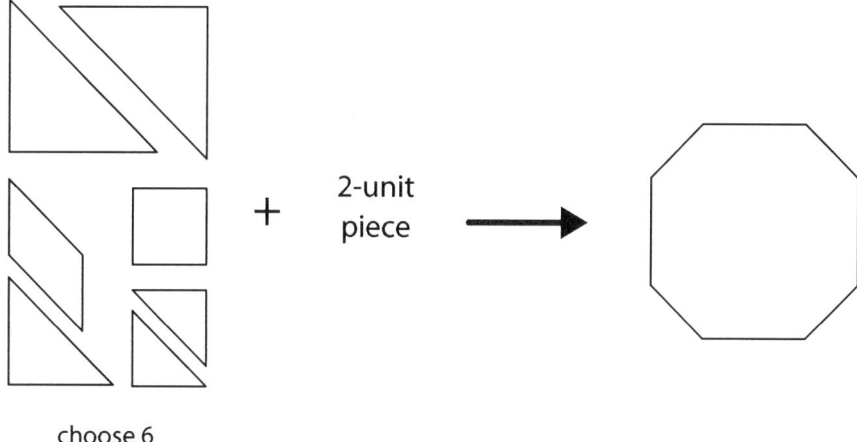

choose 6

Choose six pieces from a set and a two-unit piece from another set to create an octagon.

3.11. *The T puzzle*

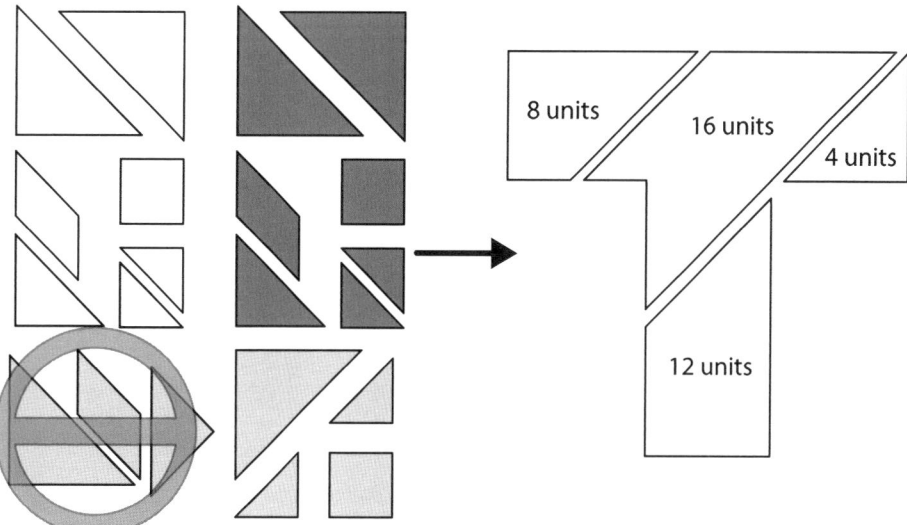

Use the pieces given above to solve the famous T-shape puzzle.

3.12.   *The M Puzzle*

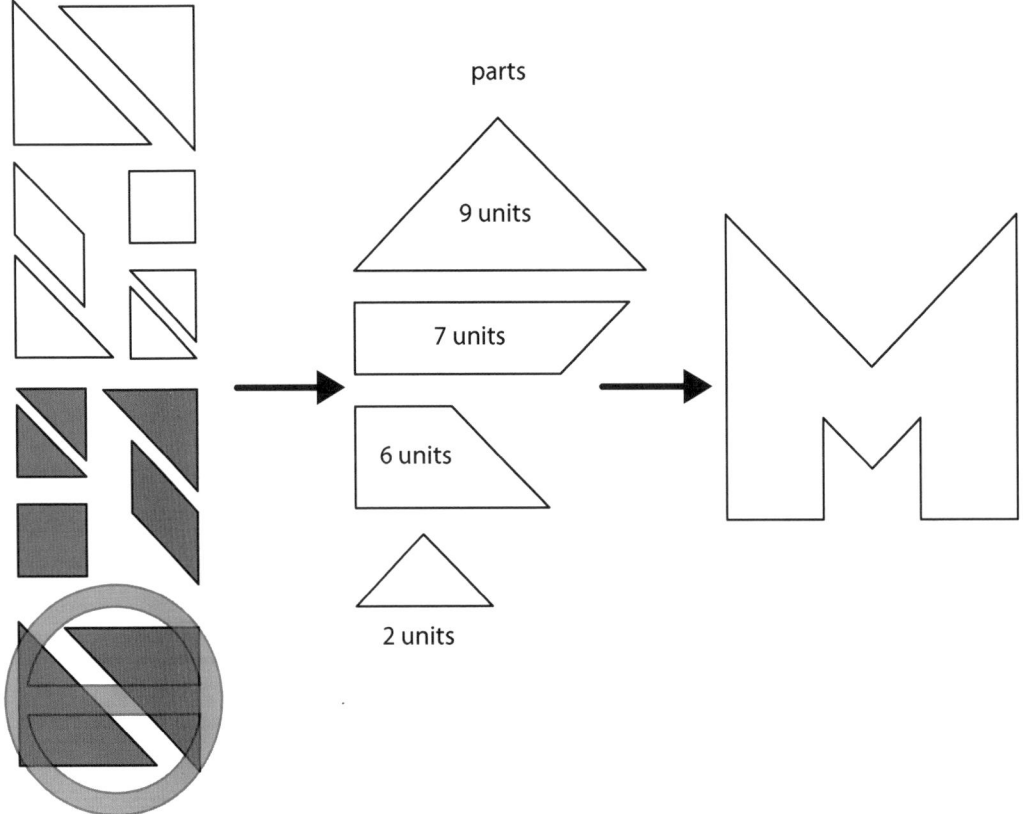

Step A: Use a set and all the small pieces of another to create the four parts of the M shape.

Step B: Solve the M-shape puzzle!

3.13.   *The N Puzzle*

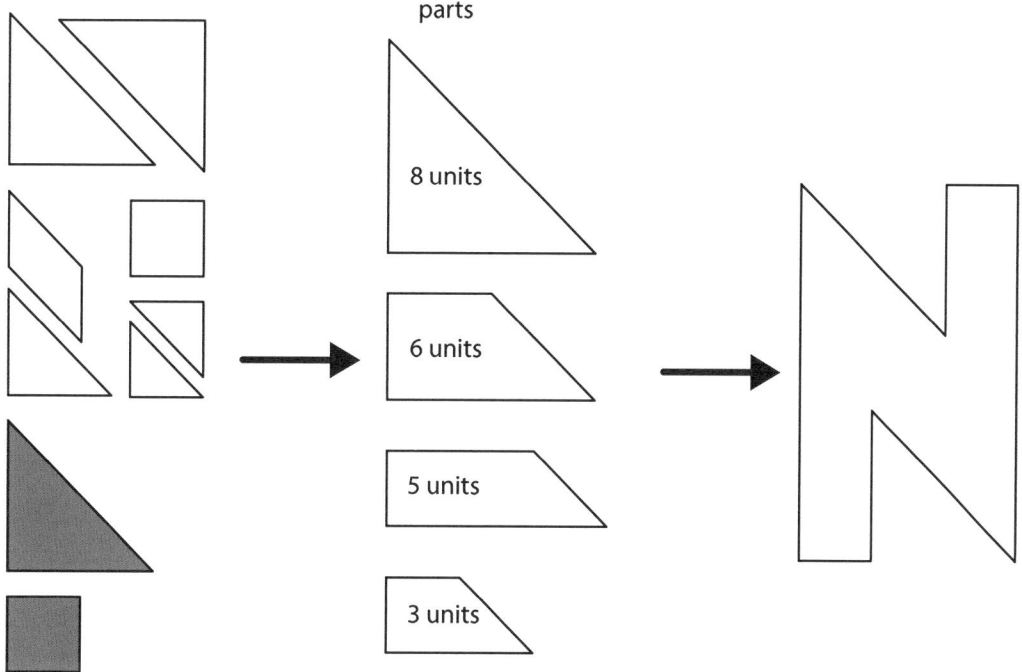

Step A: Use a set and the two pieces from another to create the four parts
of the N shape.

Step B: Solve the N-shape puzzle!

3.14. *The Pythagoras Theorem*

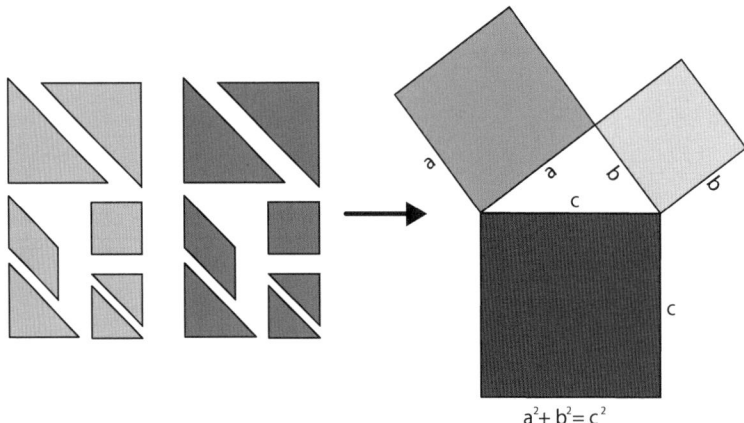

$$a^2 + b^2 = c^2$$

Use pieces from two sets to show that $a^2 + b^2 = c^2$. There are several solutions.

3.15. *The Inheritance*

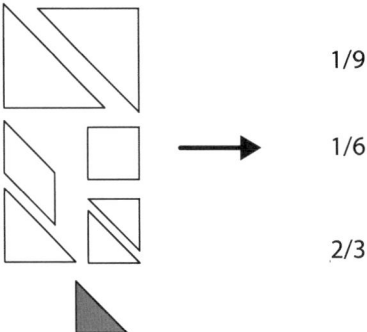

1/9

1/6

2/3

A father wants to give his Tangram treasure (consisting of a full set and a small triangle) to his three sons.

The eldest is to get two-thirds of the total area (of 17 units). The middle son is to get a sixth of it, while the youngest is to get a ninth only.

Yet, everyone must get one triangle and at least one two-unit part.

How can it be done?

3.16. *Square-Holed Octagon*

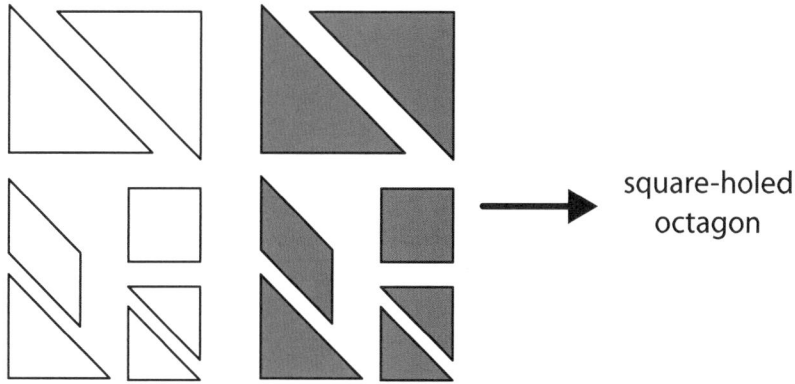

Use two sets to create an octagon with a square hole inside.

3.17. *The Fourth Division*

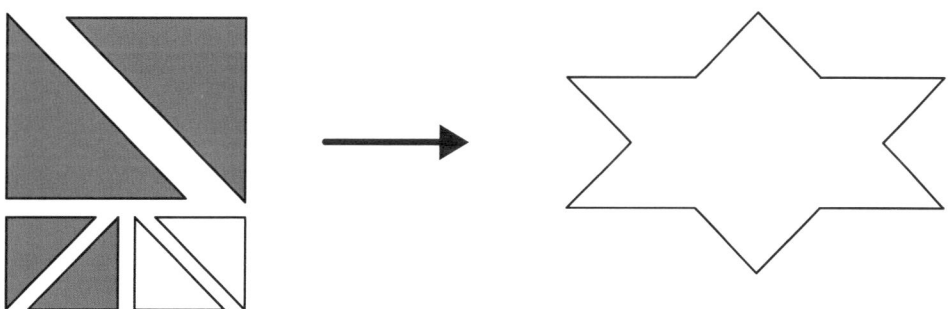

Step A: Create the six-pointed star with the pieces given above.
Step B: Divide this shape into four identical parts.

3.18. *The K Puzzle*

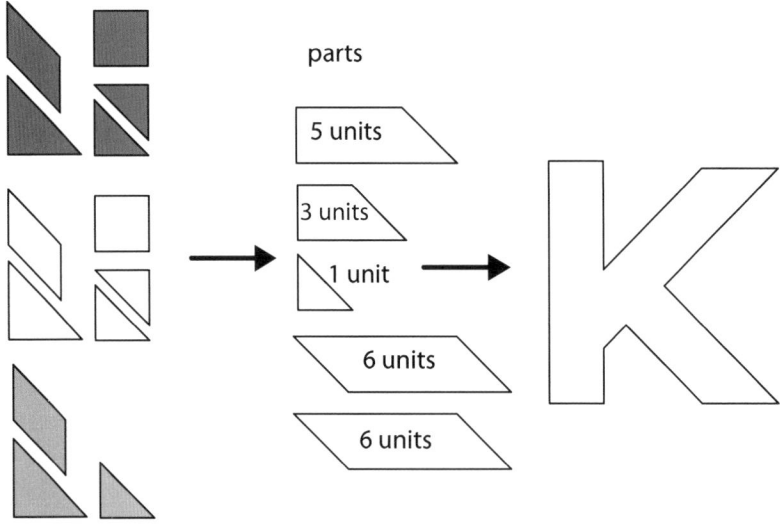

Step A: Use any of the pieces given above to create the five parts.
Step B: Use the five parts to solve the K puzzle!

3.19. *The Symmetric Duo*

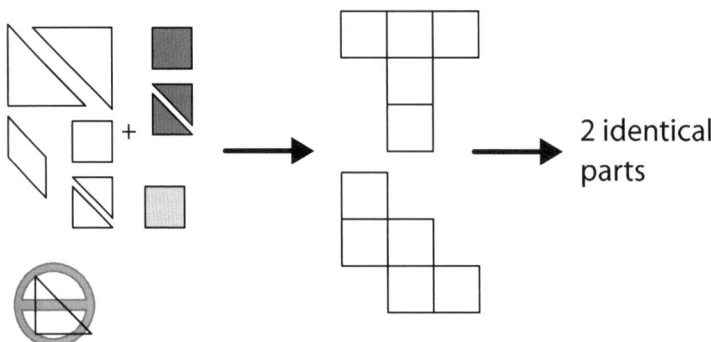

Step A: Use the above pieces from three sets to create the two pentomino shapes.
Step B: With those two shapes, create a new shape that can be divided into two identical parts.

3.20. *The Divided Rectangle*

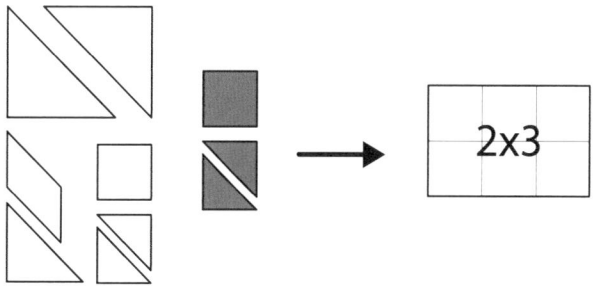

Step A: Create a 2×3 rectangle using any of the pieces shown above.
Step B: Divide it into two identical parts.

There are at least four solutions.

3.21. *A Symmetric Trick*

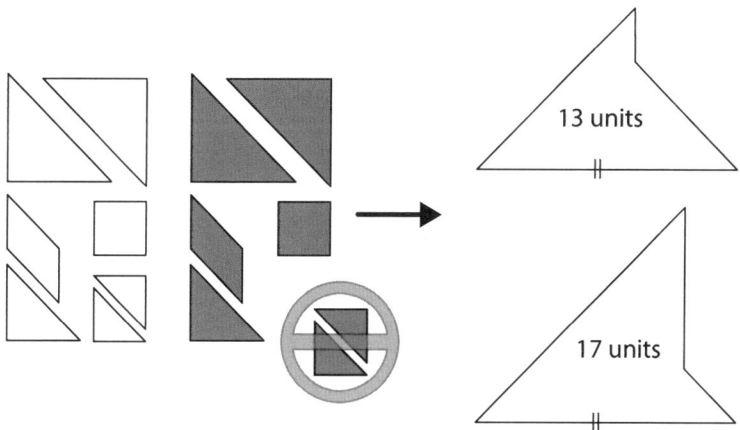

Note: The bases are of the same length.

Step A: Create the two shapes with the pieces of two sets.
Step B: Combine the two shapes together to form a symmetrical shape.

This is the Tangram version of the SymmeTRICK puzzle, by **Vesa Timonen**, Finland.

3.22.   *The CL Puzzle*

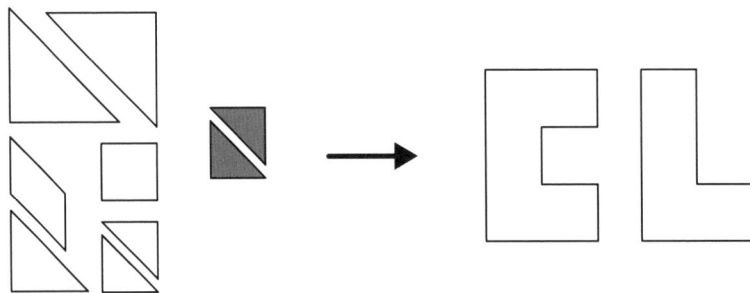

Step A: Create the letters C and L with the pieces given above.
Step B: Arrange the C and the L to form a symmetrical shape.

3.23.   *Self Similarity (Part 1 – The Triangle)*

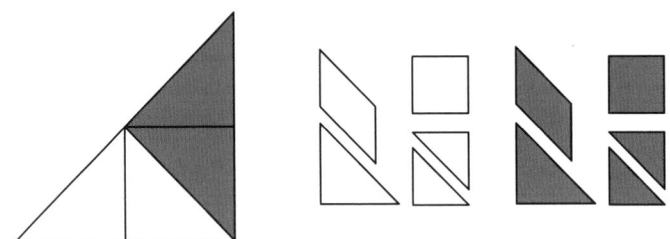

Four large triangles can be combined to form a larger triangle, of the same proportion, that is four times larger in scale than a single large triangle.

Step A: With the remaining smaller pieces, create four different parts (each part made up of four units) that will cover the larger triangle.
Step B: Recreate the shape of each part, in the same proportion but of a four-time larger scale, by combining all four parts.

Note: As we started this puzzle with a triangle, one of the four parts must be a triangle (of four units).

3.24.   *Tetraboloes and a Square (Part 1)*

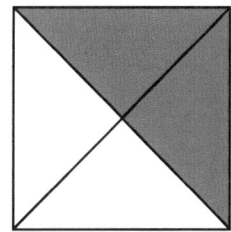   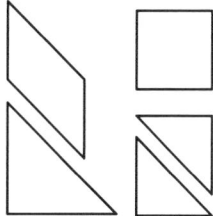

Divide the four-triangle square into four <u>identical</u> parts. Each part must be a tetrabolo that is created using the smaller pieces of unlimited sets.

Find four different solutions.

3.25.   *Tetraboloes and a Square (Part 2)*

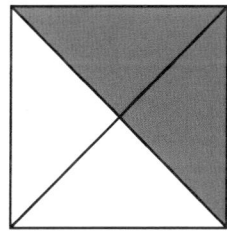   

Divide the four-triangle square into four <u>different</u> parts. Each part must be a tetrabolo that is created using the smaller pieces of unlimited sets.

Find two different solutions.

3.26. *Self Similarity (Part 2 – The Rectangle)*

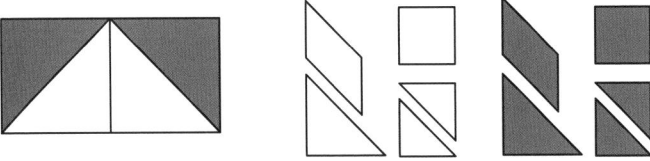

Four large triangles can be combined to form a rectangle that is four times larger in size than a single large triangle.

Step A: With the remaining smaller pieces, create four different parts that will cover the big rectangle.

Step B: Recreate the shape of each part, in the same proportion but of a four-time larger scale, by combining all four parts.

Note: As we started this puzzle with a rectangle, one of the four parts must be a rectangle.

3.27. *The Cover Up*

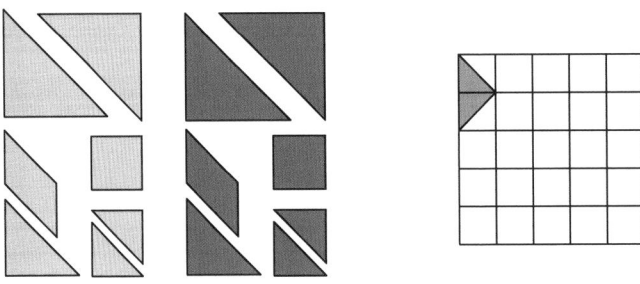

Cover as much of the 5×5 grid as possible, using the pieces of two sets. What is the maximum area you can cover?

Note: The area of a square is the same as that of the small triangle, i.e. one unit!

3.28. *Equal Rights*

Using a continuous line, divide this shape into two identical parts.

3.29. *The Bat Paradox*

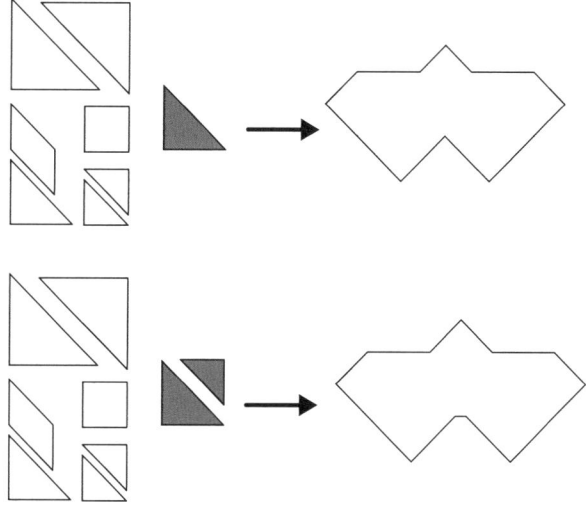

Use eight pieces to create the upper Bat logo; add another piece to create the bottom logo.

Note: In this puzzle, the pieces need not align fully along their edges.

3.30. *The Half Pyramid*

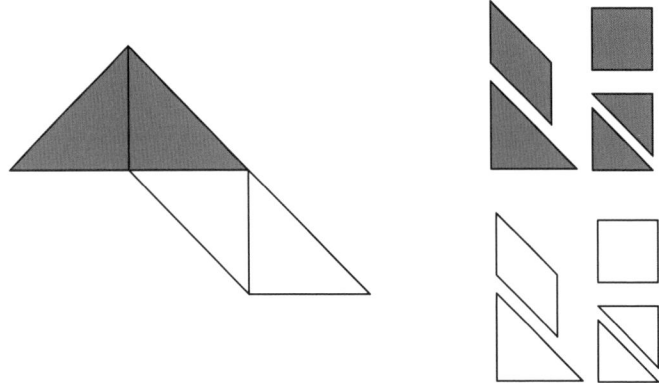

The four large triangles can be combined to form the shape of a half pyramid. Create this same shape with four different tetraboloes, using the remaining pieces of two sets.

3.31. *The Powerful Three*

There are only four possible triboloes (shapes that consist of three right-angled isosceles triangles). Above, we created the four triboloes using small pieces only.

Try and create, as in the equations below, each of the large triboloes with the four smaller triboloes.

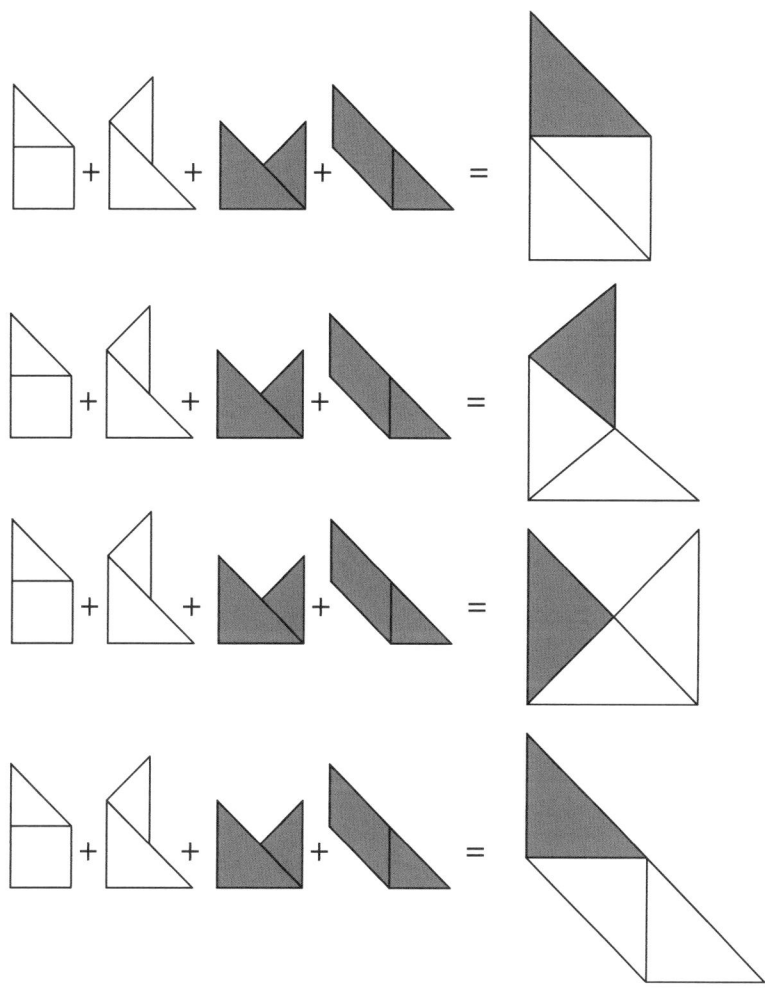

One of the equations is impossible. Find which one.

3.32. *The Mega Cover Up*

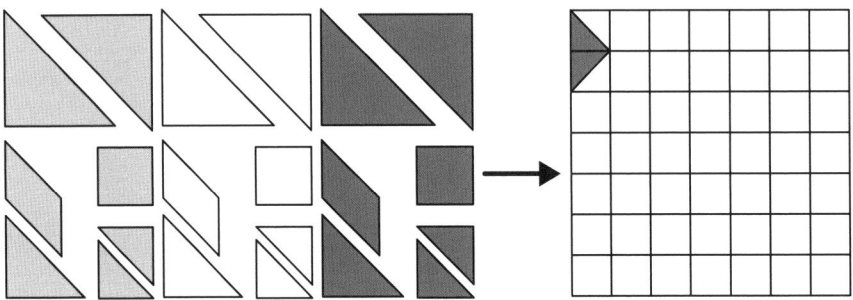

Using three sets, cover as much of the 7×7 grid as possible.

Note: The area of a square is the same as that of the small triangle, i.e. one unit.

3.33. *The Chair*

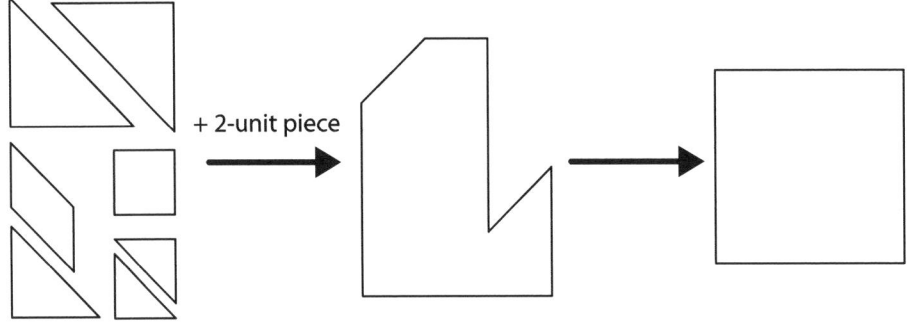

Step A: Create the shape in the middle, by using a set and an additional two-unit piece.

Step B: Divide the shape into two parts, and rearrange them to form a square.

Note: For this puzzle, a piece cannot be divided into units!

3.34.   *Self Similarity (Part 3 – The Trapezium)*

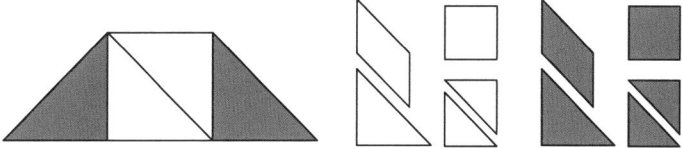

Four large triangles can be arranged to form a 16-unit trapezium.

Step A: With the remaining smaller pieces, create four different parts that will cover the large trapezium.

Step B: Recreate the shape of each part, in the same proportion but of a four-time larger scale, by combining all four parts.

Note: As we started this puzzle with a trapezium, one of the four parts must be a trapezium.

One of the shapes is impossible to solve. Find which one.

3.35.   *The Boat Paradox*

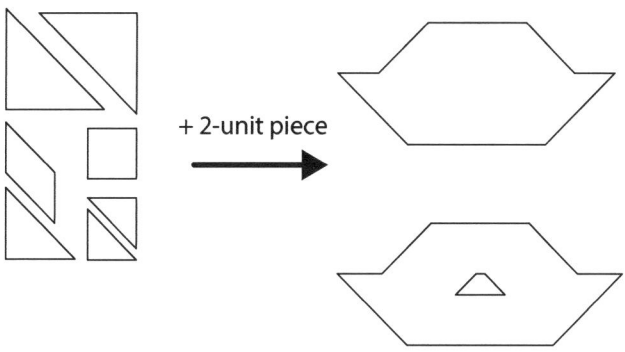

Use a set and a two-unit piece to create both boats.

Note: You can use different two-unit pieces for each of the boats.

3.36.   *The Four-Square Symmetry*

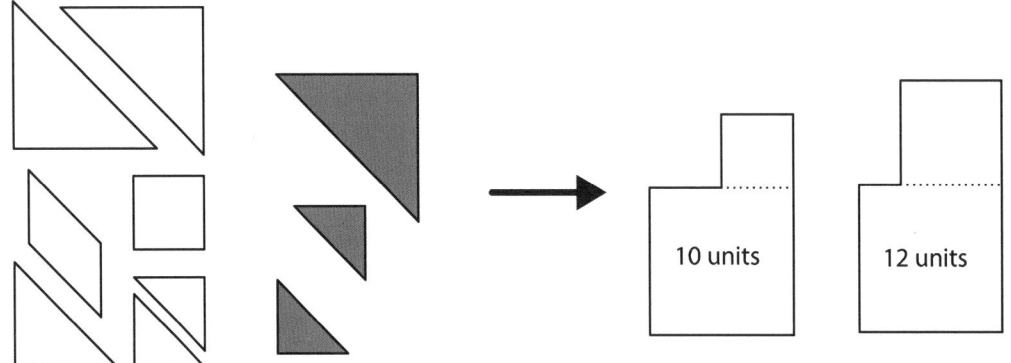

Step A: Create the shapes above with the given pieces. You can see that each shape can be divided into two squares (along the dotted lines).

Step B: Rearrange the two shapes to form a symmetrical shape.

3.37.   *Looking Back*

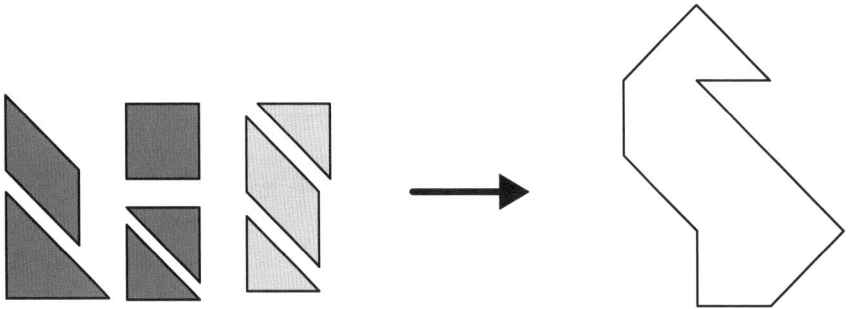

Step A: Create the shape above with the given pieces.

Step B: Divide the shape into two identical parts.

3.38. *Which Way?*

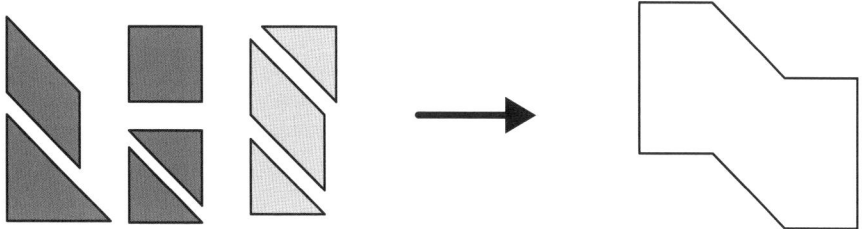

Step A: Create the shape above with the given pieces.
Step B: Divide the shape into two identical parts.

There are at least four solutions.

Note: For this puzzle, a piece cannot be divided into units!

3.39. *The Stop Sign*

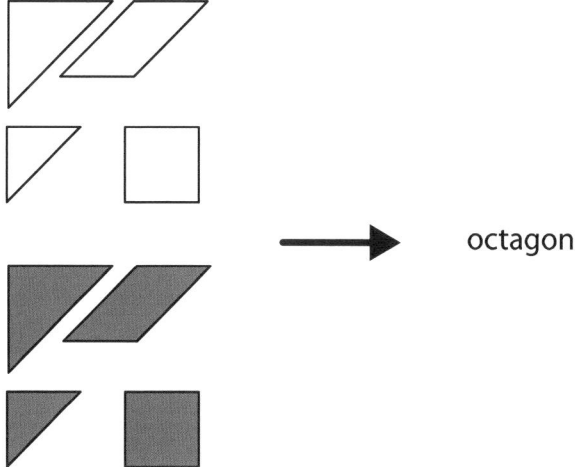

octagon

Using only the pieces given above, create an octagon.

Try and find only symmetrical solutions.

3.40.   *All Different!*

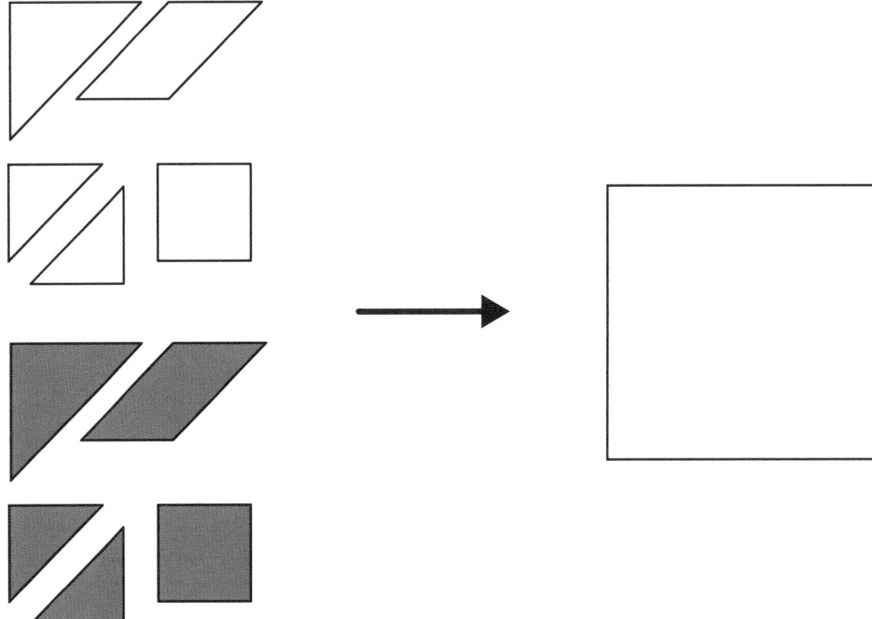

Using only the pieces given above, create a square made up of four triboloes and two 2-unit pieces.

The triboloes must all be different!

Solutions

3.1. *The Square Deal*

Step A (two solutions)

Step B

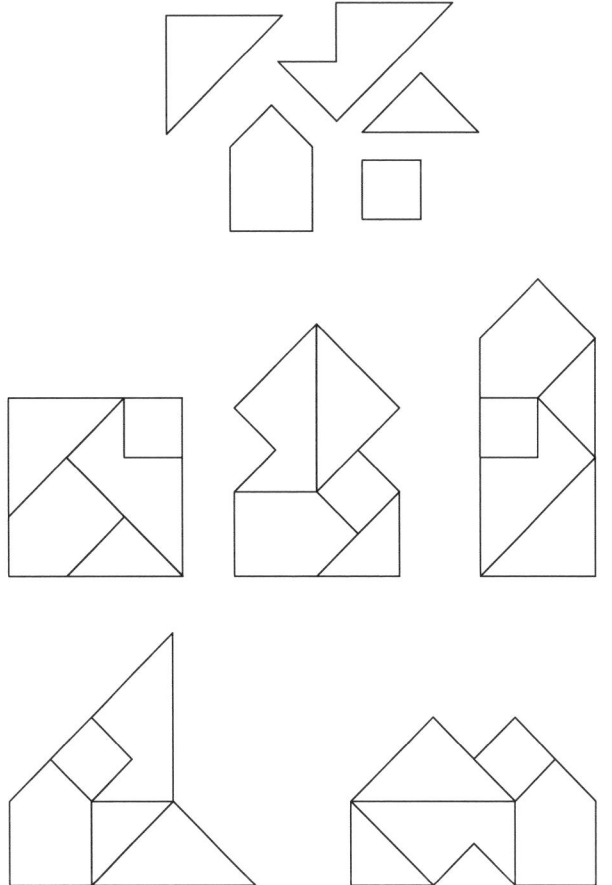

3.2. XL Six-Pointed Star

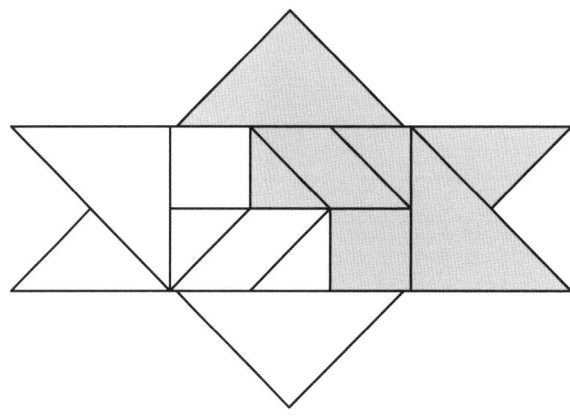

3.3. The XL Square

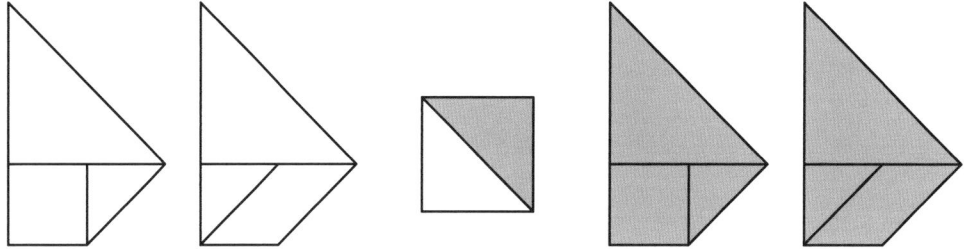

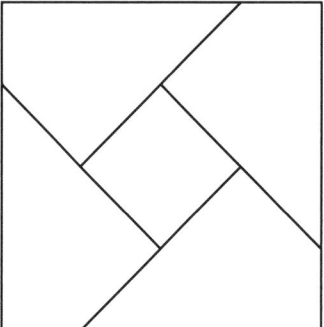

3.4. *Max Spinners*

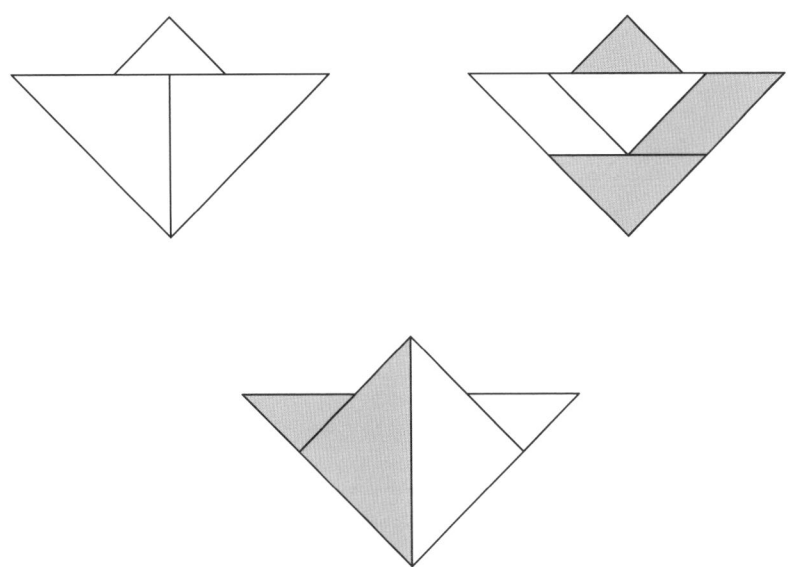

3.5. *The Pentomino Rectangle*

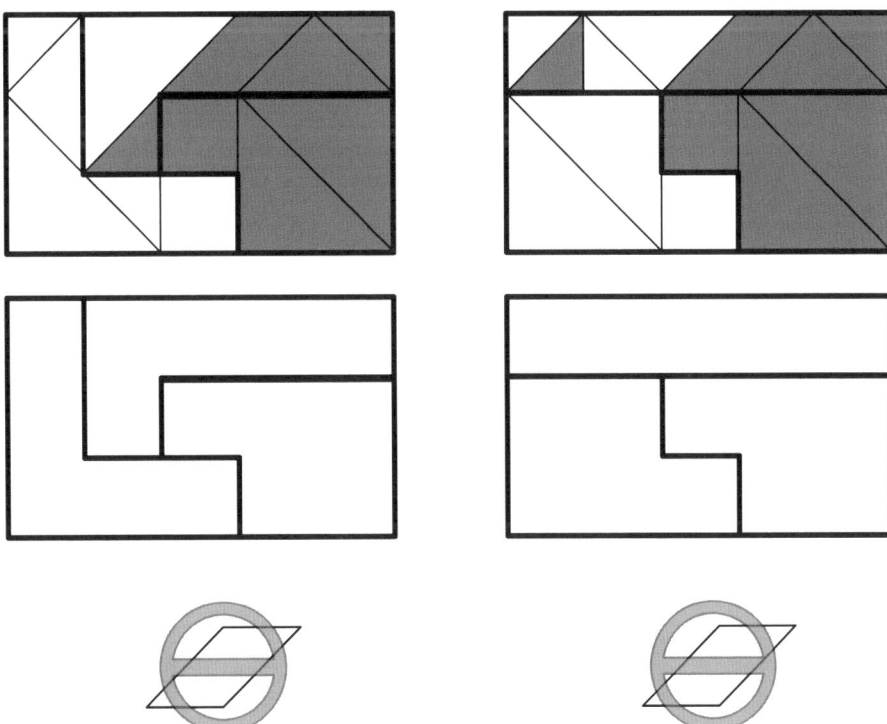

3.6.   *The Largest Pentomino Rectangle*

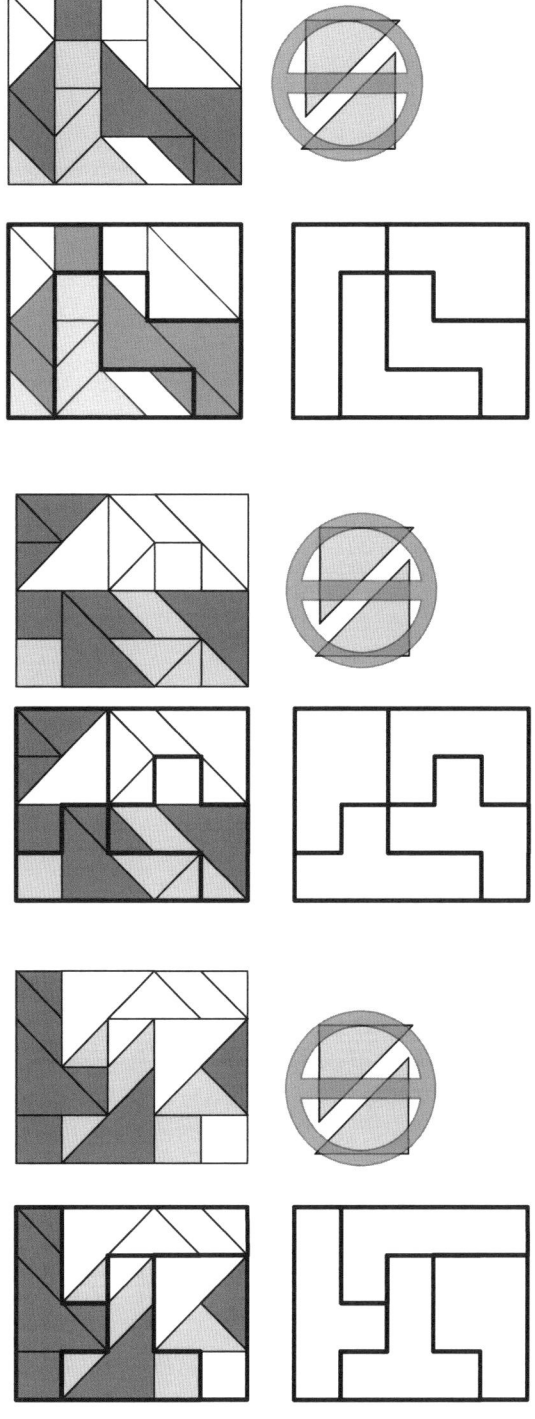

3.7. *The Parrot*

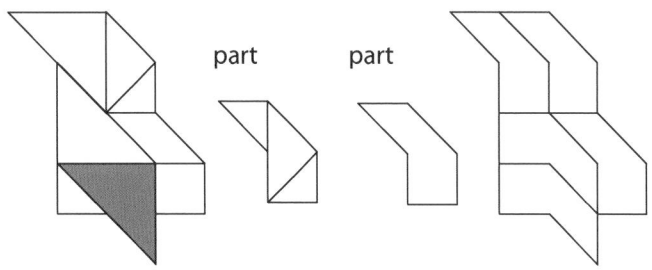

part   part

3.8. *Square and Octagon*

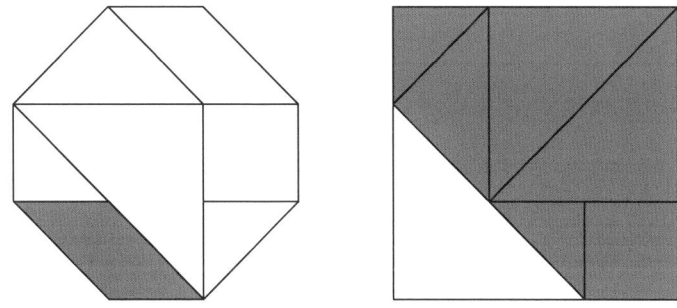

3.9. *Five, Six, Seven*

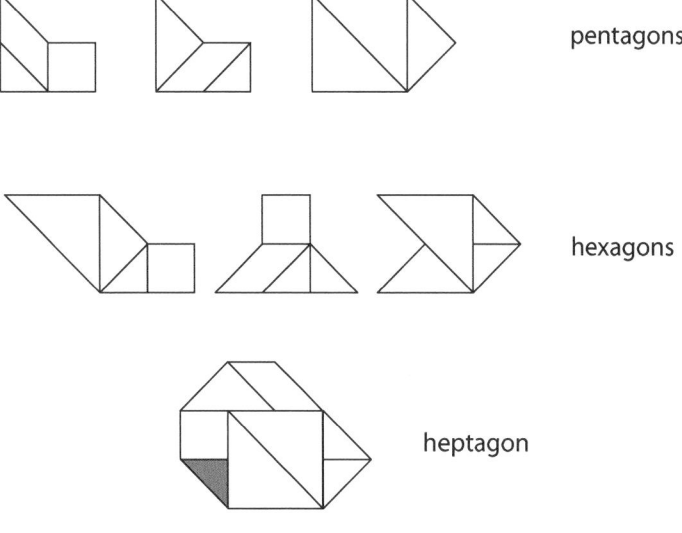

pentagons

hexagons

heptagon

3.10. *Seven to Eight*

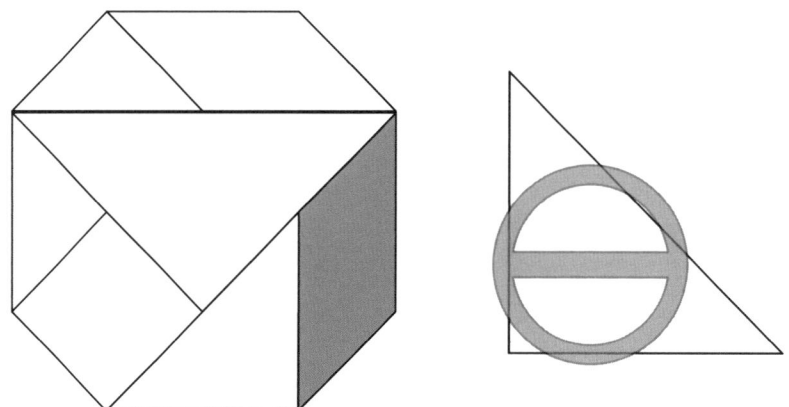

3.11. *The T puzzle*

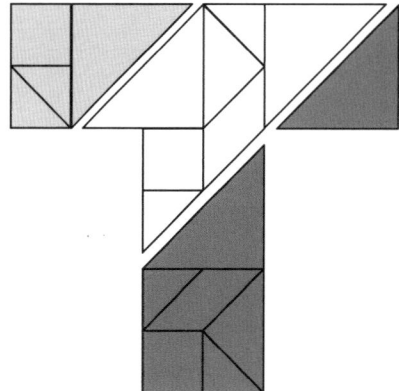

3.12. *The M Puzzle*

3.13. *The N Puzzle*

3.14. *The Pythagoras Theorem*

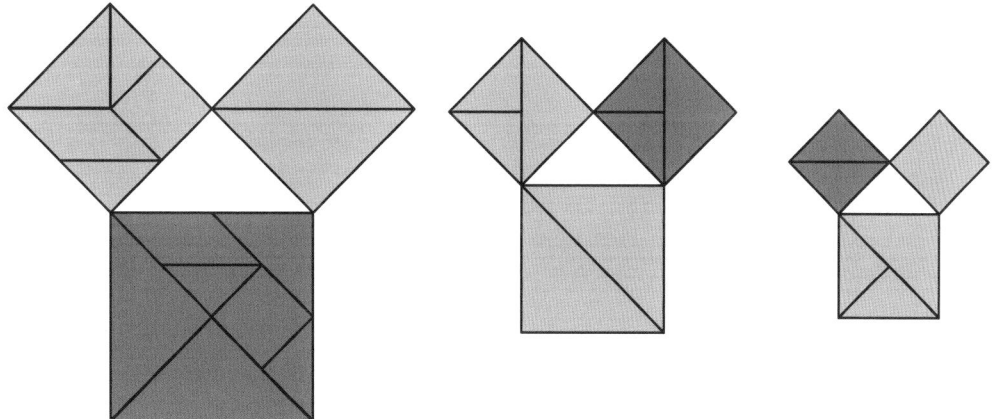

3.15.   *The Inheritance*

Like the famous puzzle this is based on, we must add a unit to solve this puzzle. By doing that we get 18 units to divide, so the eldest son gets twelve $(18 \times \frac{2}{3})$, the middle son gets three $(18 \times \frac{1}{6})$ and the youngest son gets two $(18 \times \frac{1}{9})$. The total sum is $12 + 3 + 2 = 17$, so the extra piece can be returned.

Bearing in mind the conditions, here are two solutions:

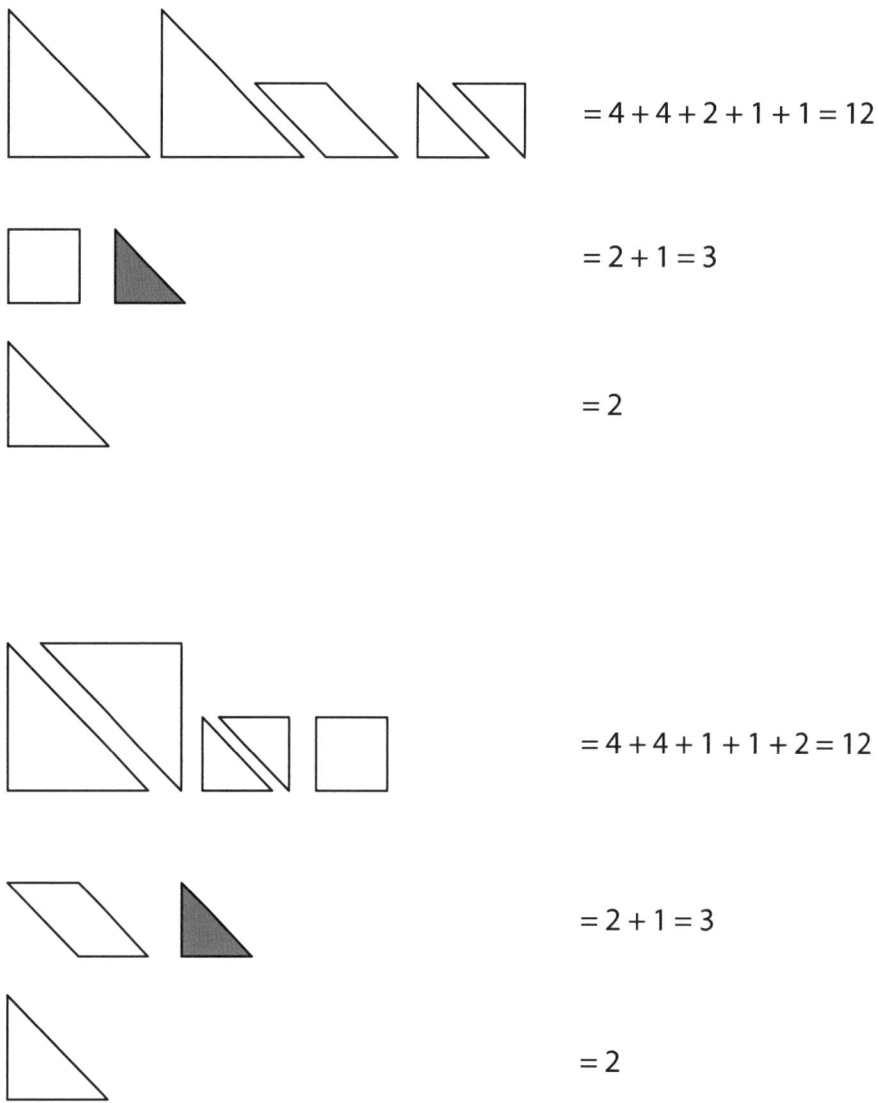

3.16. *Square-Holed Octagon*

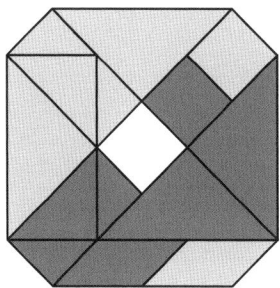

3.17. *The Fourth Division*

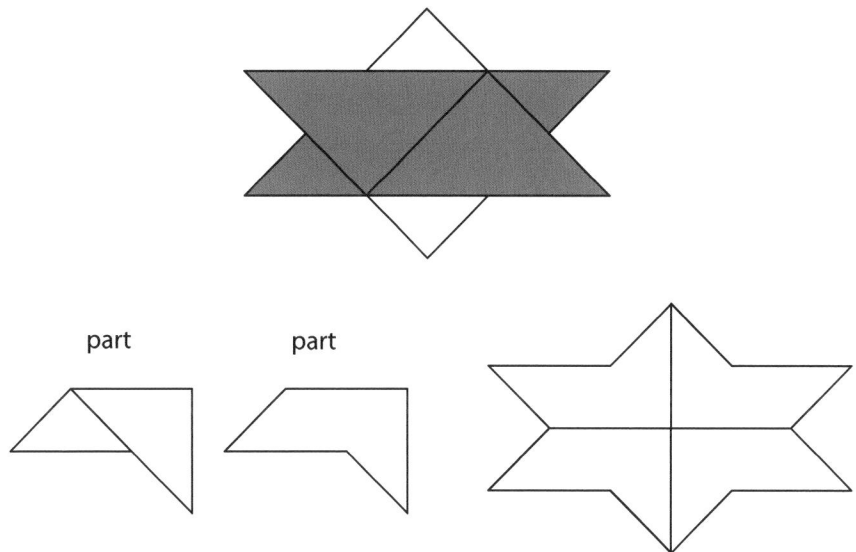

part   part

3.18. *The K Puzzle*

3.19.   *The Symmetric Duo*

Step A

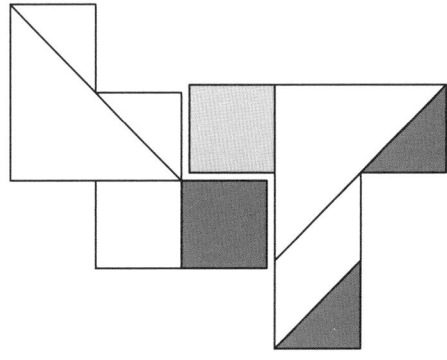

Step B

3.20.   *The Divided Rectangle*

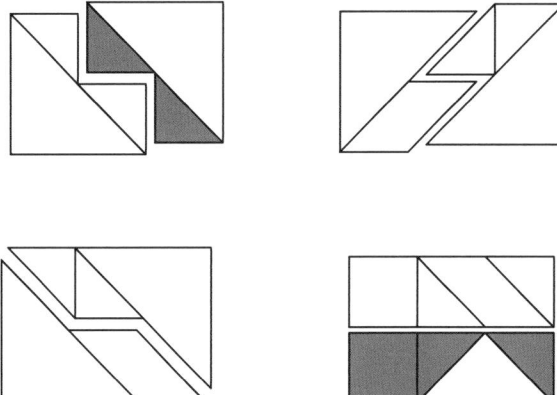

3.21.   *A Symmetric Trick*

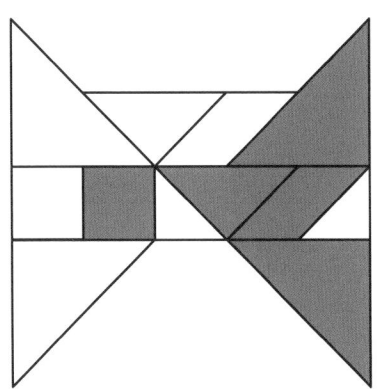 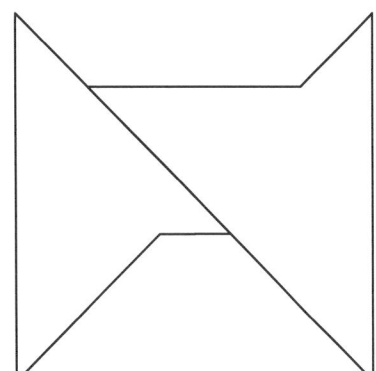

3.22.   *The CL Puzzle*

 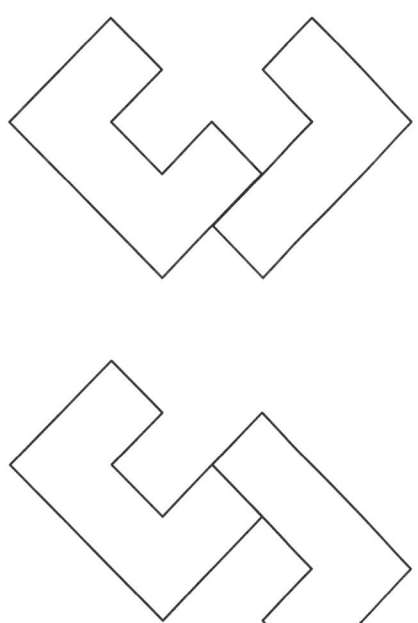

3.23.   *Self Similarity (Part 1 – The Triangle)*

part   four-time larger version of the part

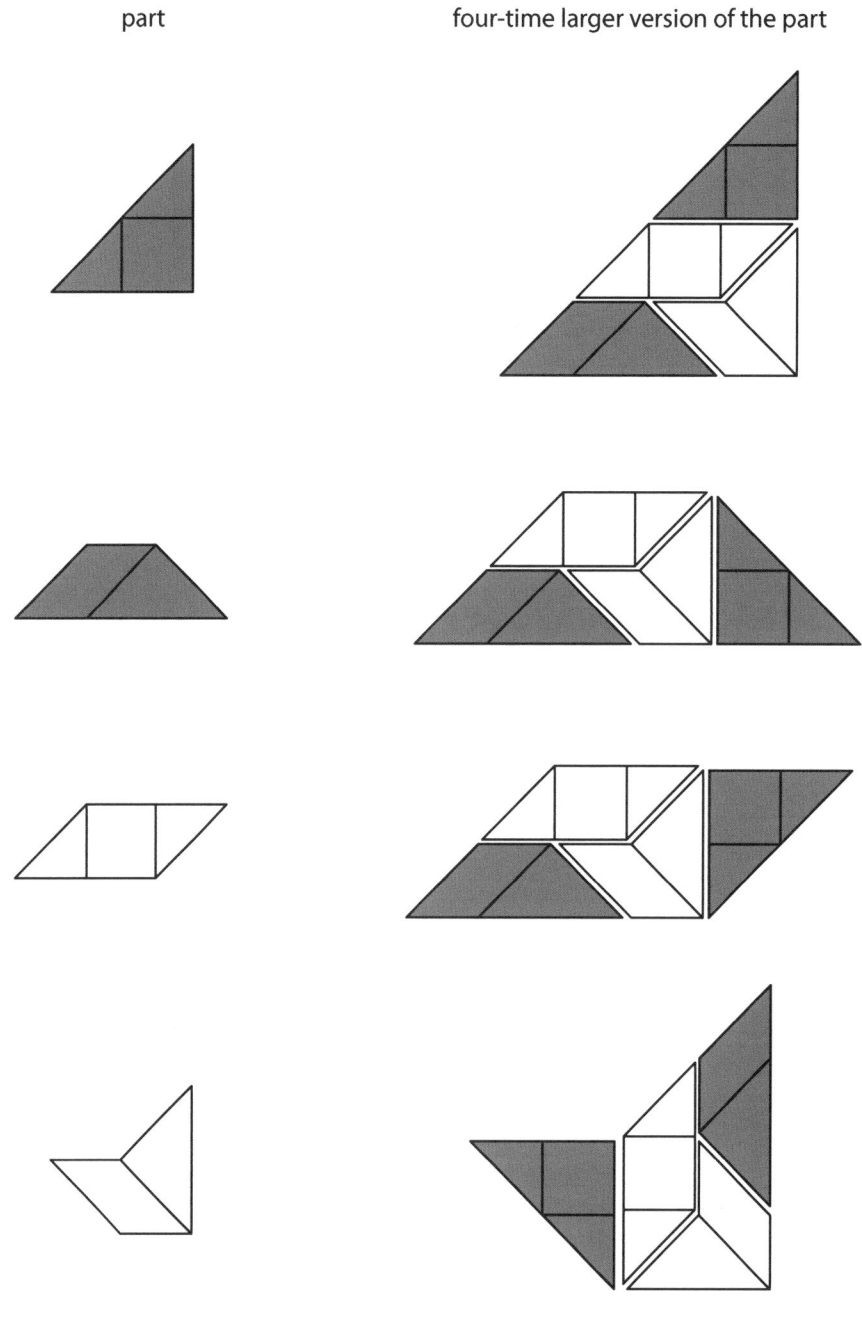

3.24.   *Tetraboloes and a Square (Part 1)*

tetrabolo part

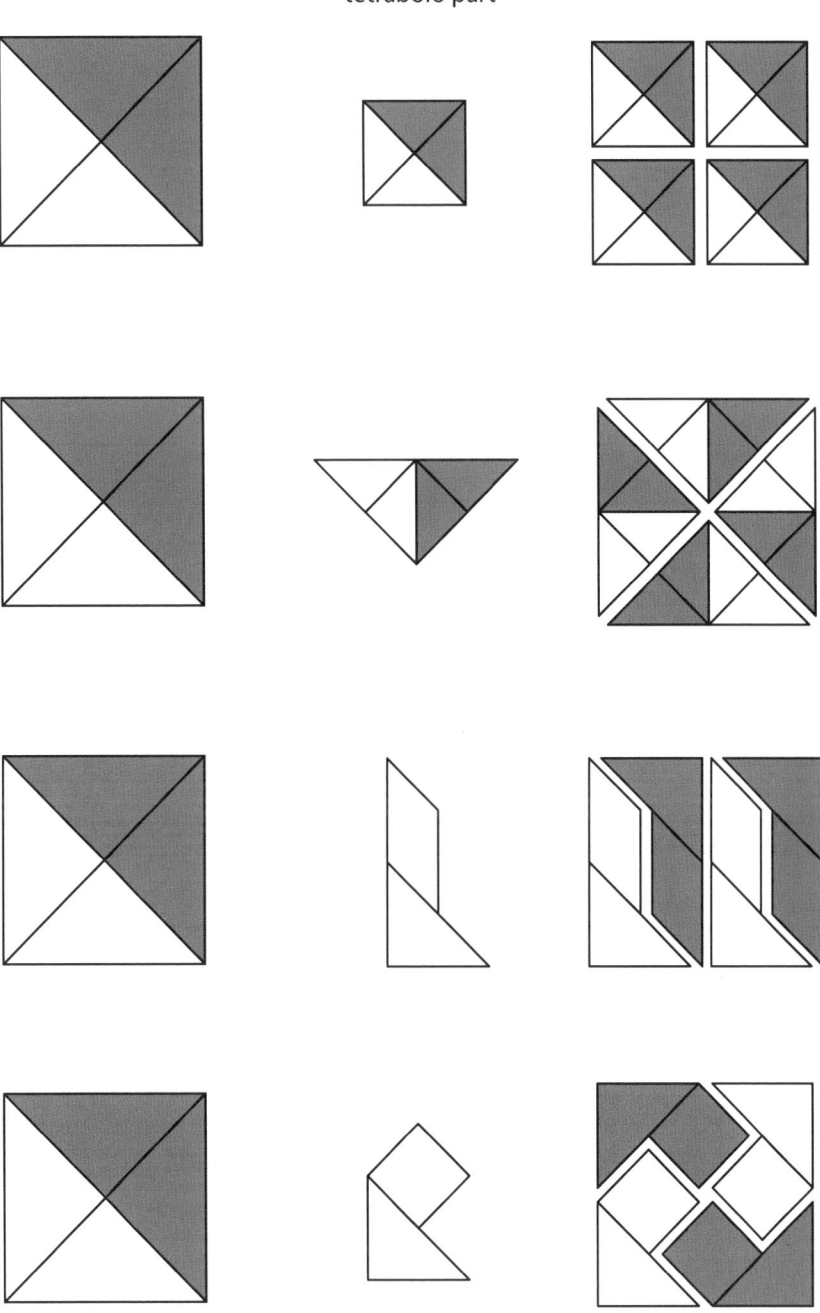

3.25.   *Tetraboloes and a Square (Part 2)*

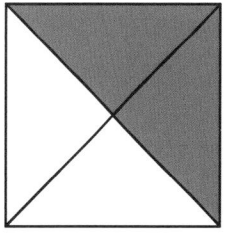  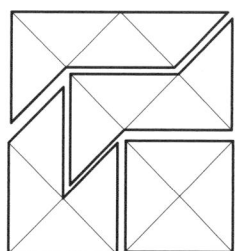

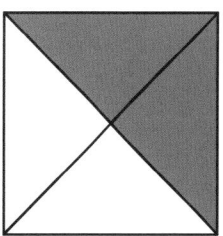  

3.26. *Self Similarity (Part 2 – The Rectangle)*

Step A

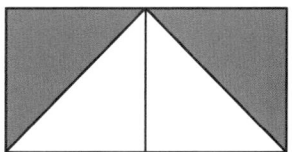

Step B

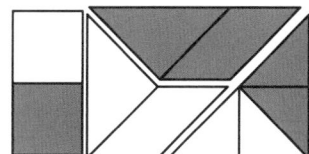

part   four-time larger version of the part

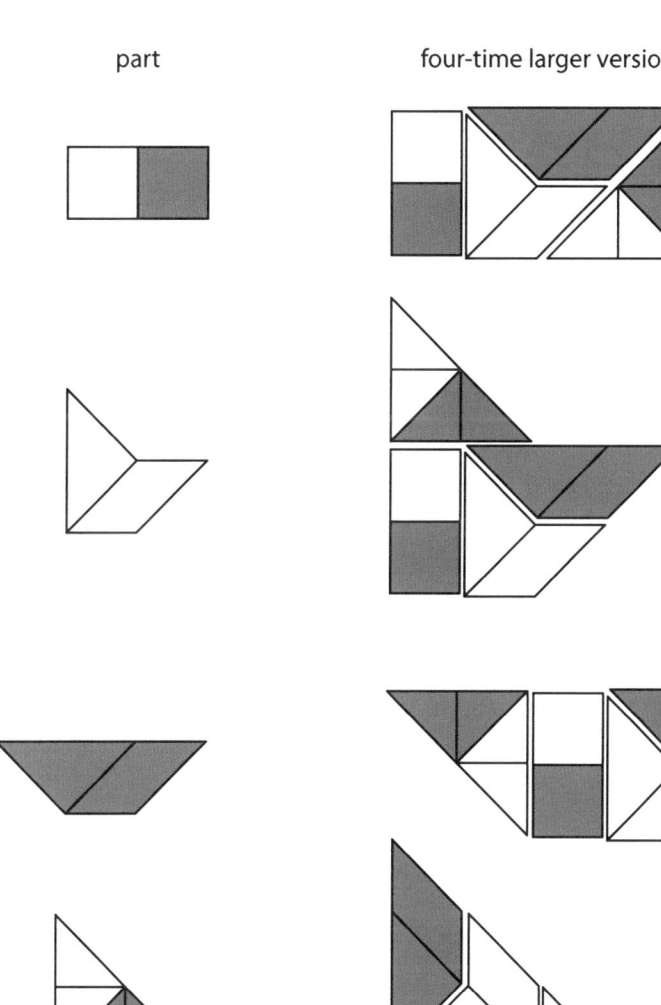

3.27.   *The Cover Up*

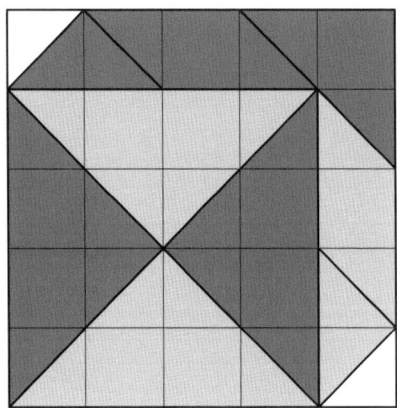

You can cover up to 24 out of the 25 squares.

3.28.   *Equal Rights*

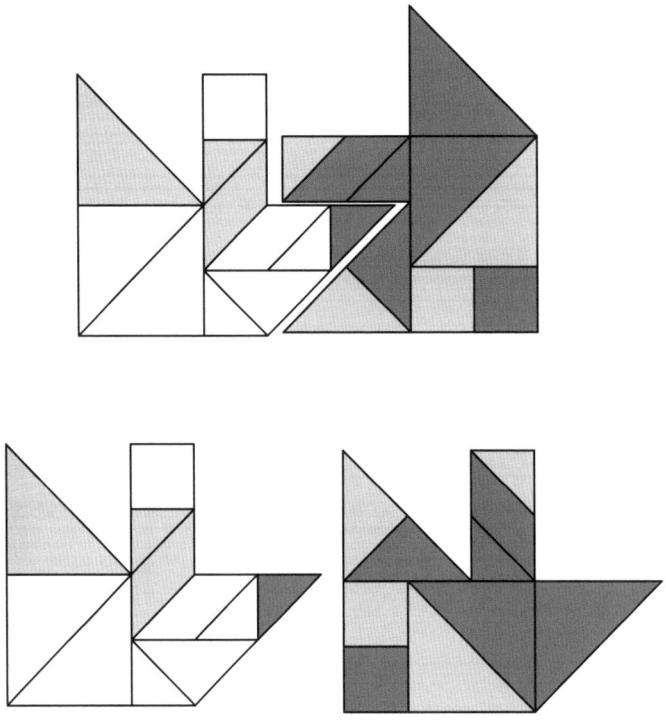

3.29. *The Bat Paradox*

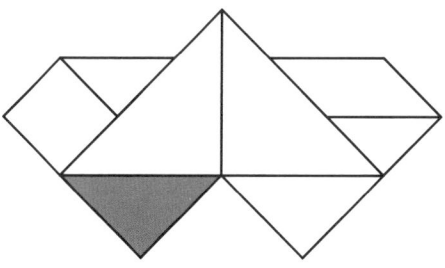

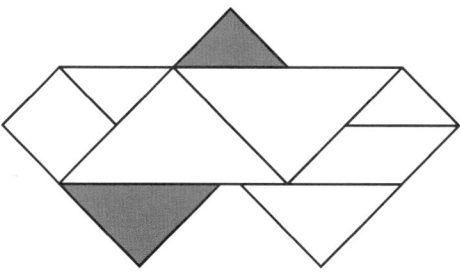

3.30. *The Half Pyramid*

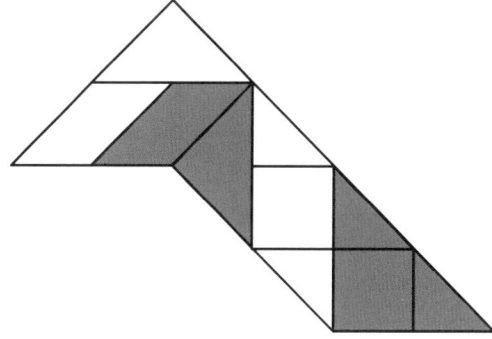

3.31. *The Powerful Three*

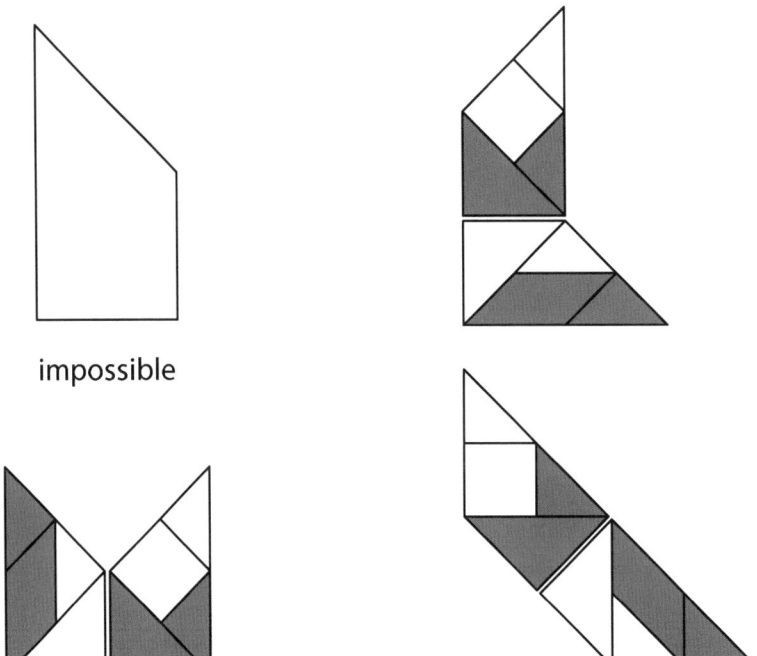

impossible

3.32. *The Mega Cover Up*

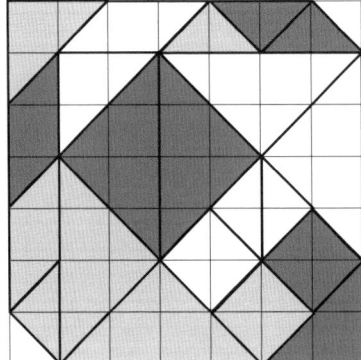

You can cover up to 48 out of the 49 squares.

3.33. *The Chair*

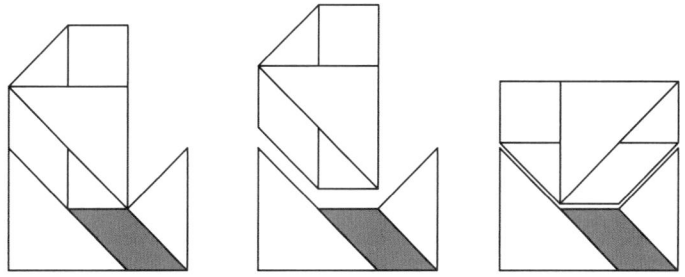

3.34. *Self Similarity (Part 3 – The Trapezium)*

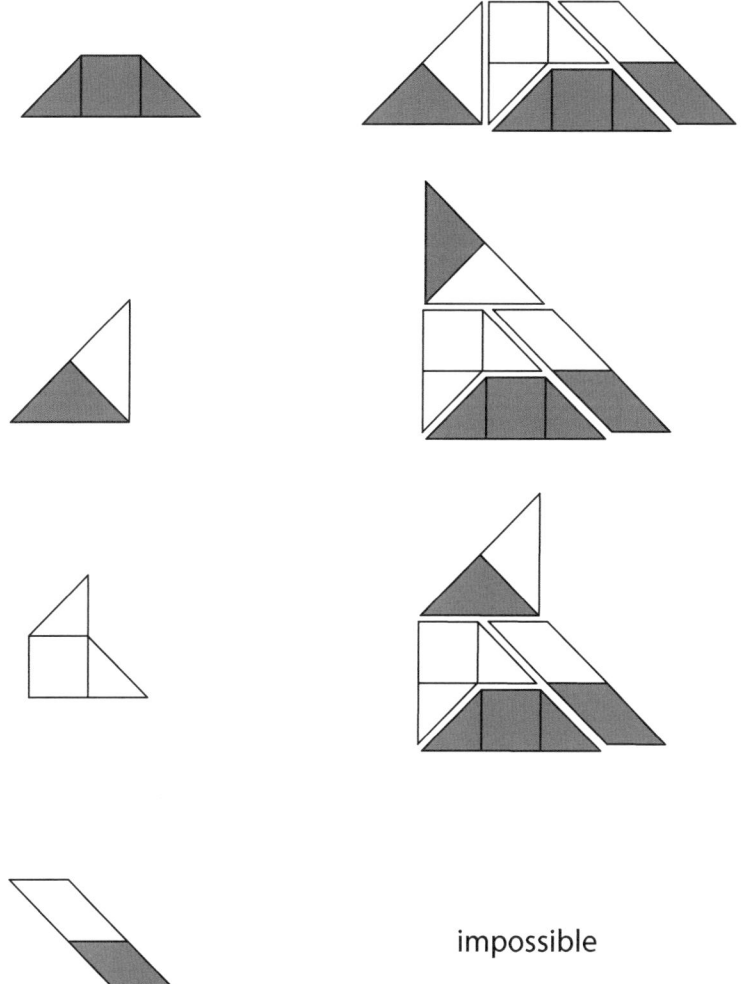

impossible

3.35. *The Boat Paradox*

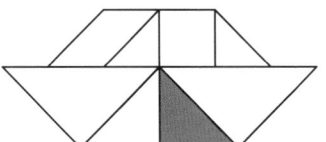

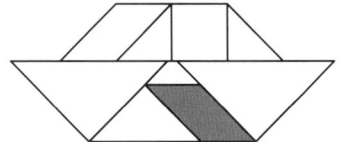

3.36. *The Four-Square Symmetry*

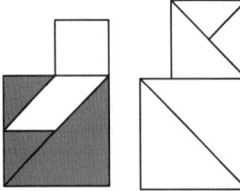

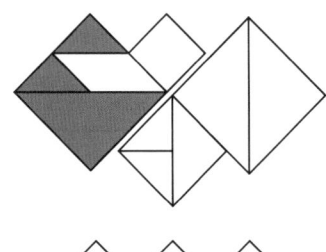

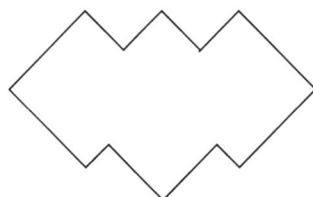

3.37. *Looking Back*

3.38.   *Which Way?*

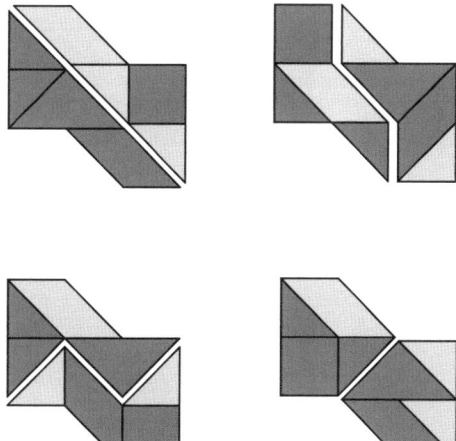

3.39.   *The Stop Sign*

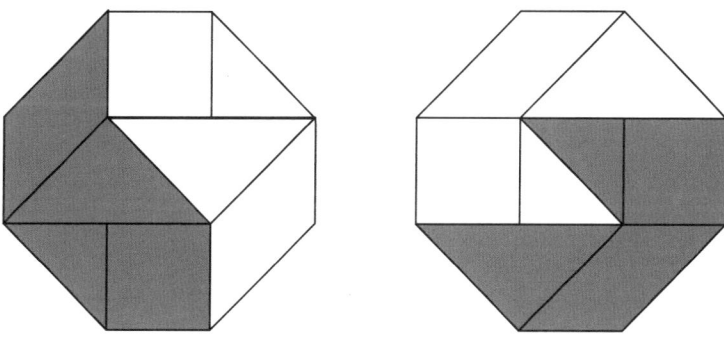

3.40.   *All Different!*

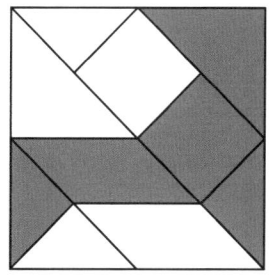 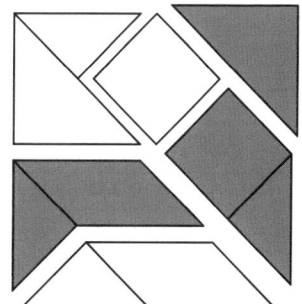

Chapter 4

Beyond the Tangram

This chapter is devoted to puzzles that follow the essence of the Tangram but do not use the actual pieces of the set.

Here, we may ask you to *break* some of the rules we set out in the beginning of this book, for example, dividing a piece into its units and overlapping the pieces!

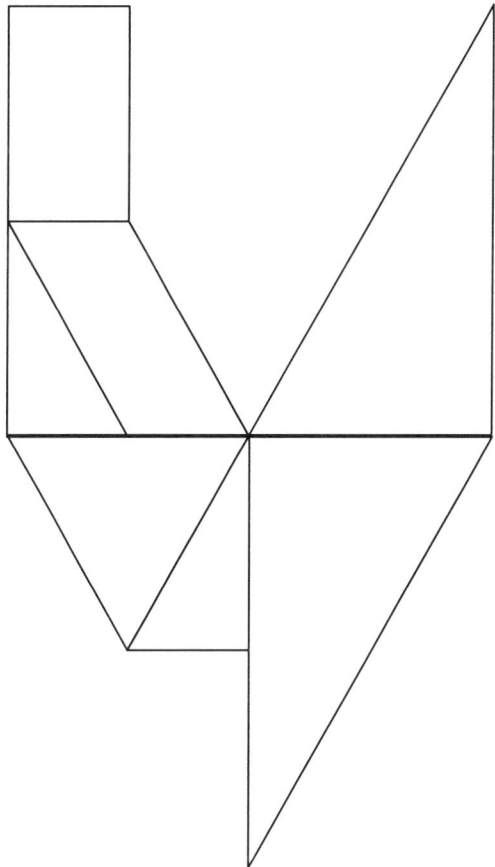

4.1.   *The 49/50 Paradox*

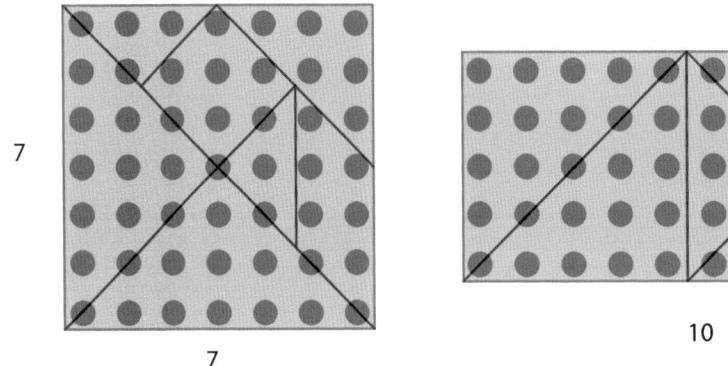

The arrangement of the pieces of a Tangram set to form a classic square covers a grid of 7×7 dots (= 49 dots); but rearranging the pieces to form two smaller squares covers only a grid of 5×10 dots (= 50 dots). 49 dots or 50 dots?

Explain this contradiction.

The Contour Tangram

Imagine that the Tangram pieces are transparent, and you only see the outlines of the pieces. By overlapping these transparent Tangram pieces, partially or fully, we can create some interesting outlines.

4.2. *Contour Tangram Puzzles*

In the example below, we can create the outline on the right by overlapping the seven transparent pieces.

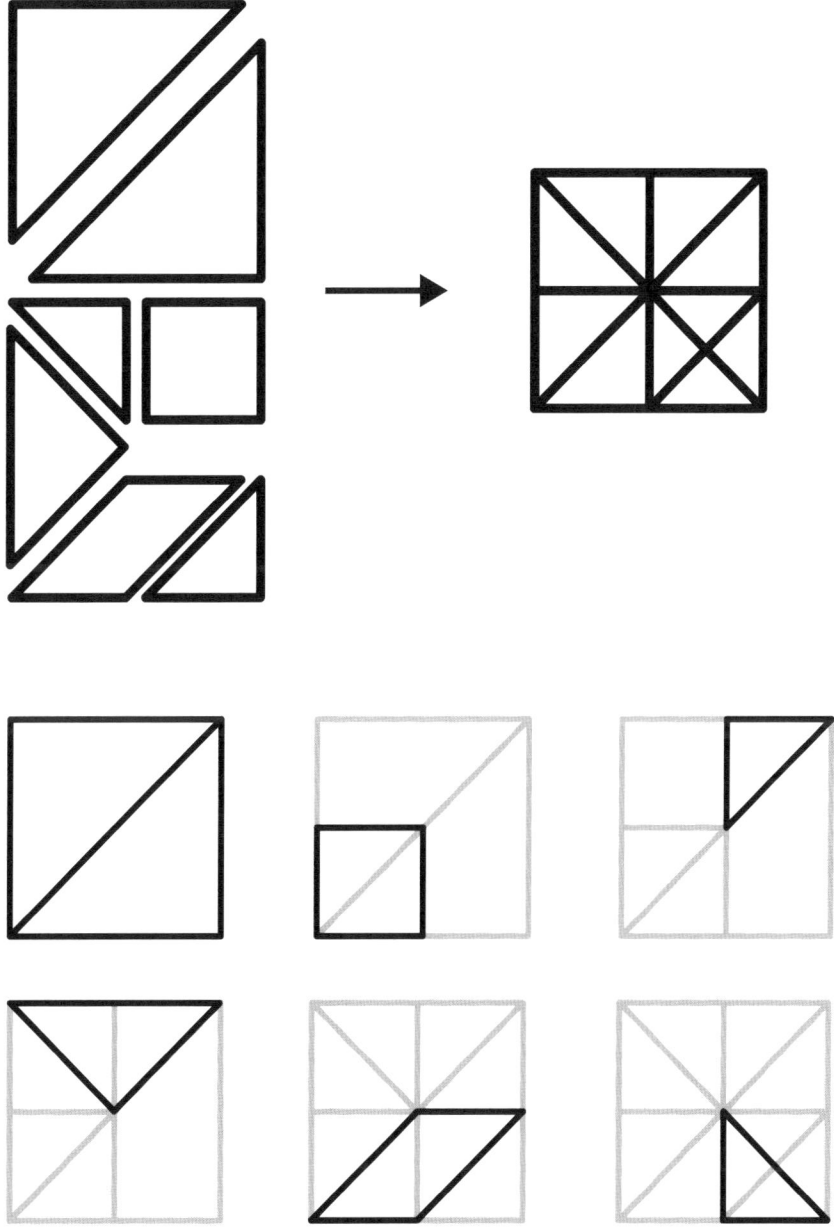

This concept was invented by **Serhiy Grabarchuk**, USA.

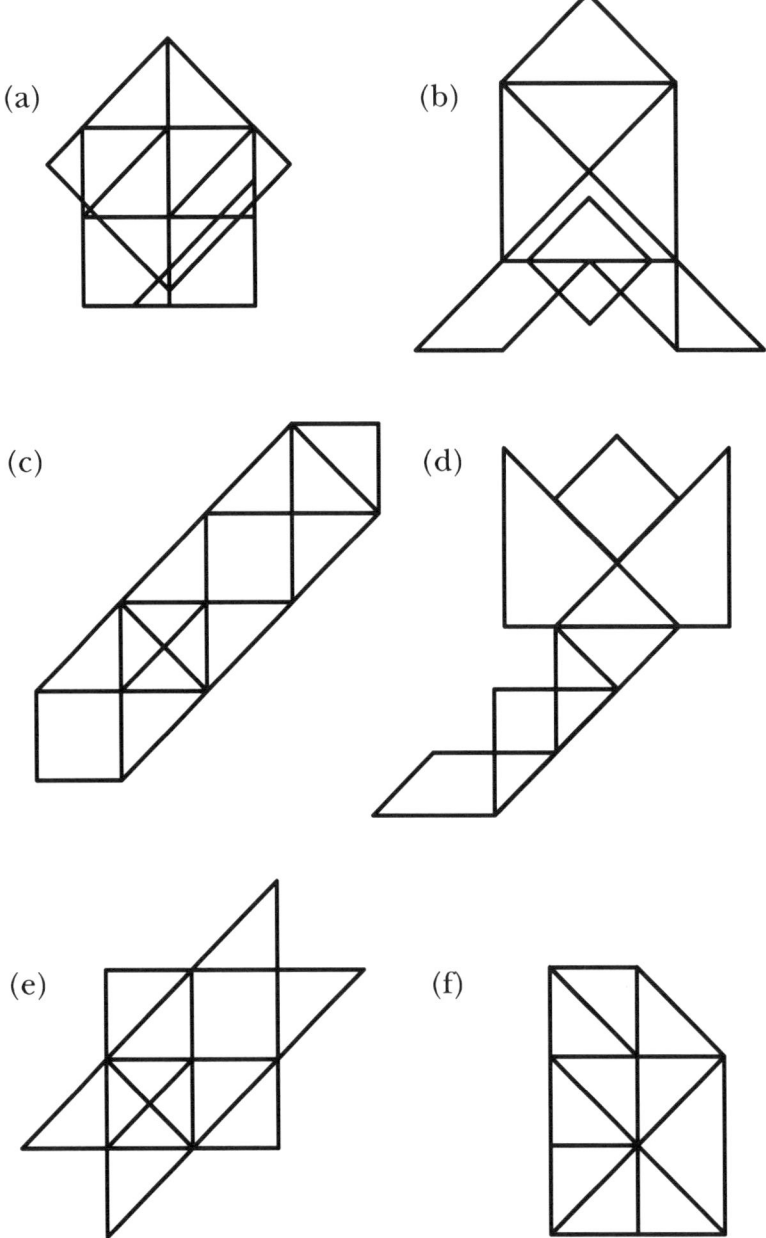

(a)   (b)

(c)   (d)

(e)   (f)

Arrange the seven pieces of a transparent Tangram set to get the outlines shown above.

4.3. *Tangram Battleship*

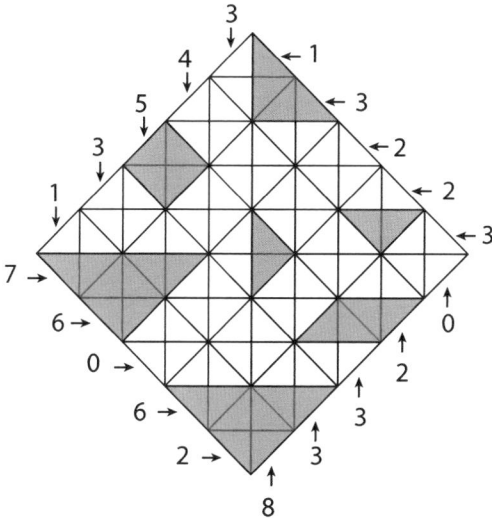

Like the classic Battleship game, each piece is considered as a vessel. The numbers represent the number of **units** in each row or column.

Your mission is to find the location of your enemy's fleet. No two vessels can have a common edge or corner.

This puzzle was invented by **Serhiy Grabarchuk**, USA.

Here are three challenges:

Fleet 1

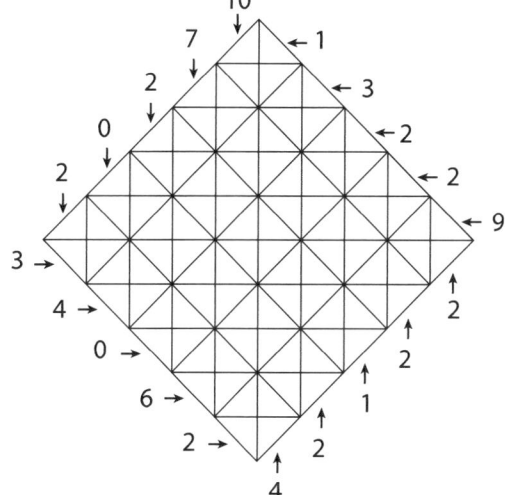

Fleet 2

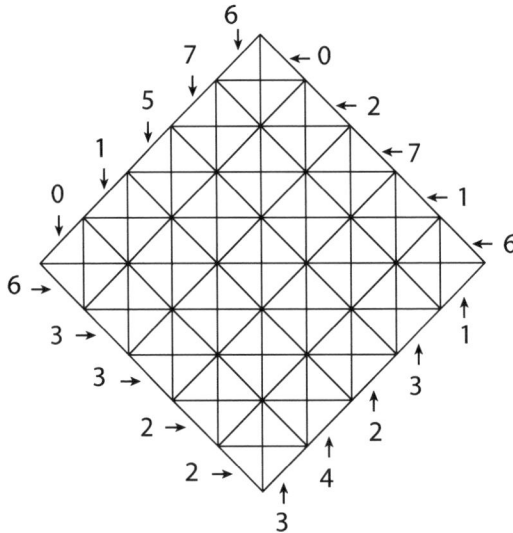

Fleet 3 – No row or column can have zero units.

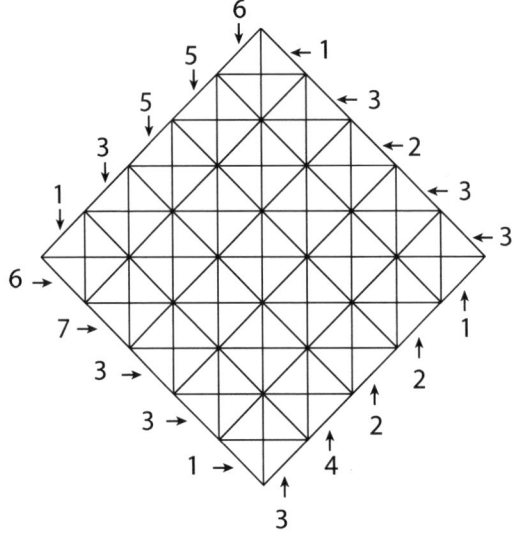

4.4.   *TanFrames: Triangles*

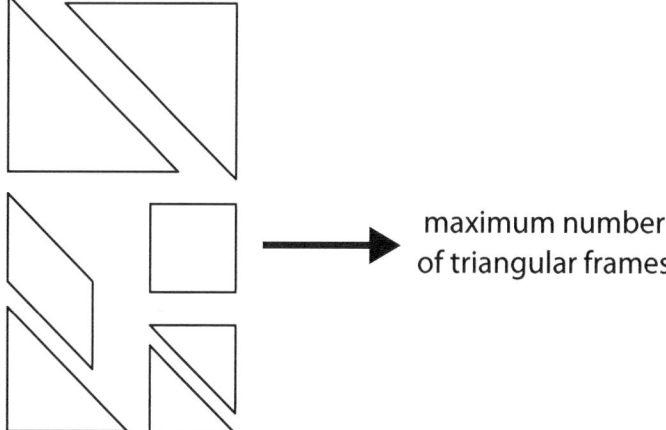

maximum number
of triangular frames

Arrange the seven pieces of a set to create as many triangular frames as possible using the outlines of the pieces.

This is a variation of a puzzle by **Serhiy Grabarchuk**, USA.

4.5.   *30:60:90 Tangram*

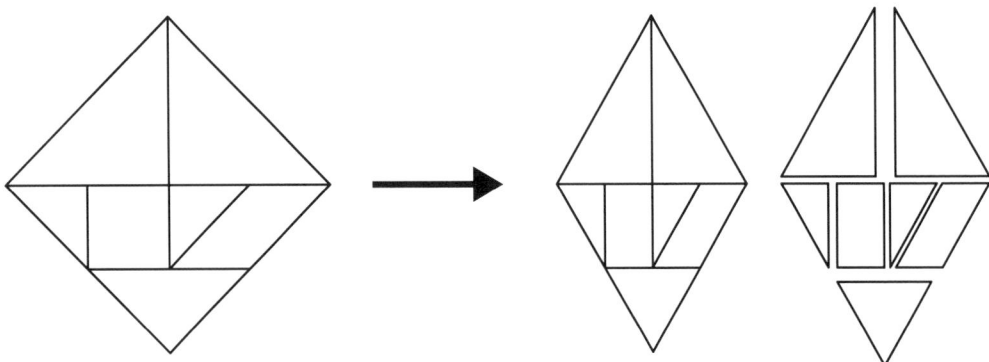

Imagine squeezing a set of Tangram pieces so that the angles of the triangles are changed from 45:45:90 to 30:60:90.

Most of the puzzles in this book can be recreated using this alternative Tangram set, and even classic Tangram silhouettes can be made more challenging this way.

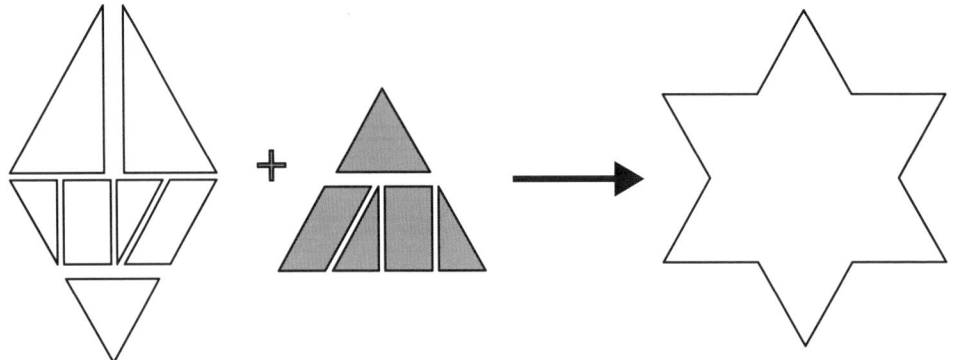

With a set and all the small pieces of another, create a perfect six-pointed star. Bonus points for a symmetrical solution!

4.6.   *30:60:90 Silhouettes*

Uncover the structure of each silhouette, using a single 30:60:90 set.

4.7. *30:60:90 Stacking Trapeziums*

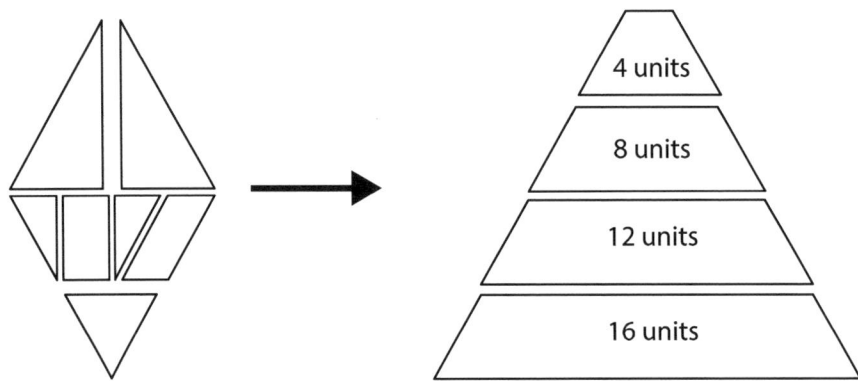

Create the base trapezium with a full set. Take out four units to create the one above it. Continue with this process till you reach the top.

4.8. *The Matching Cube*

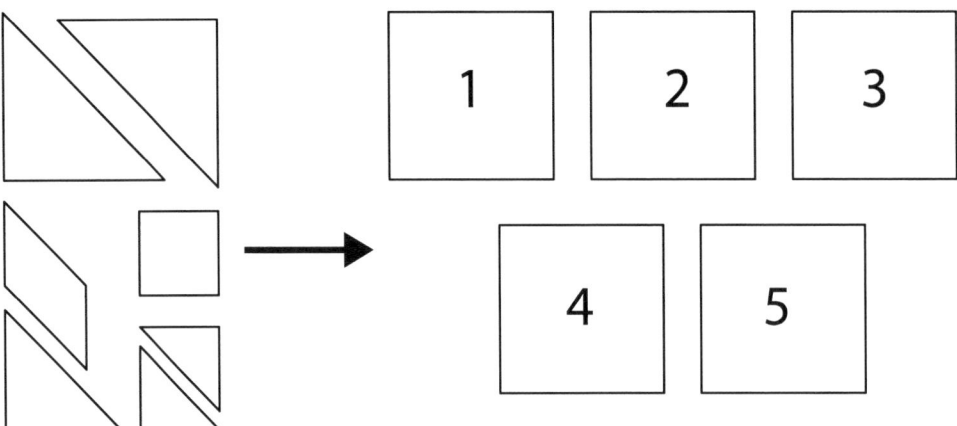

Step A: Choose pieces that have eight units in total. There are five ways to create a square with these pieces. Find them all.

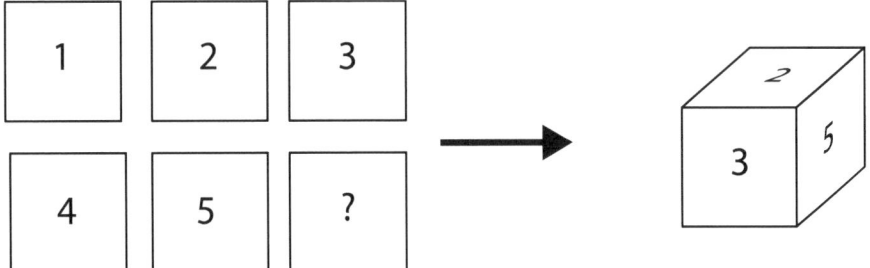

Step B: Place the five parts from Step A (one should be used twice!) on a cube, so that the edge of each *face* matches the edge of an adjacent one.

Two edges are said to be *matched* if

(a) they are both the base of a four-unit triangle, or
(b) they are both made up of two short edges (edge of a square or the base of the small triangle).

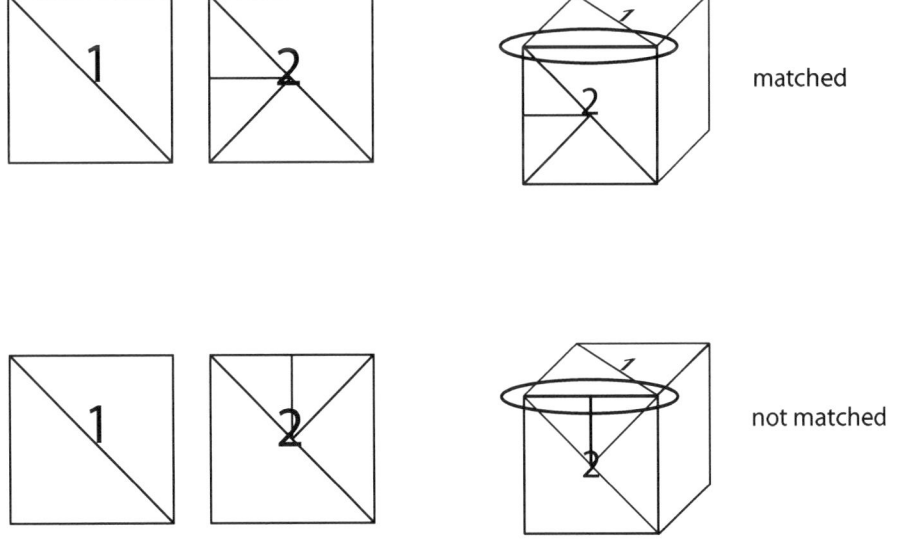

Choose wisely which part to use twice!

Chipped Puzzles

Chipped puzzles are made up of Tangram pieces that are broken into units, or are *chipped*.

4.9.   *The Square Skata*

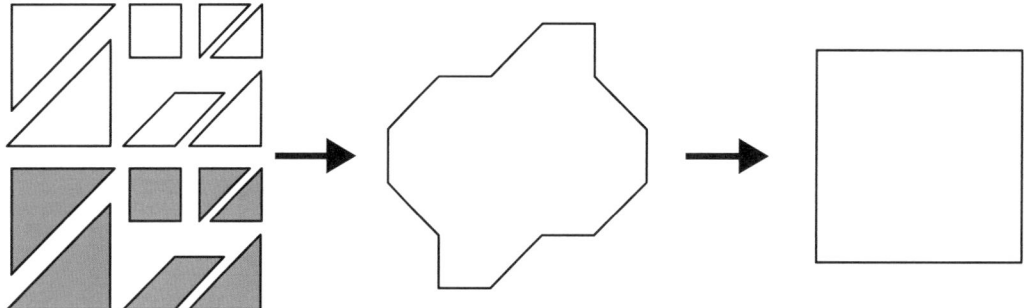

Step A: Arrange the pieces of two Tangram sets to form the shape shown in the middle above.

Step B: Divide the shape into four identical parts that can be rearranged to form a square.

This is the Tangram version of the Square Skata puzzle by **Peter Grabarchuk**, USA.

4.10. *The Fishy Division*

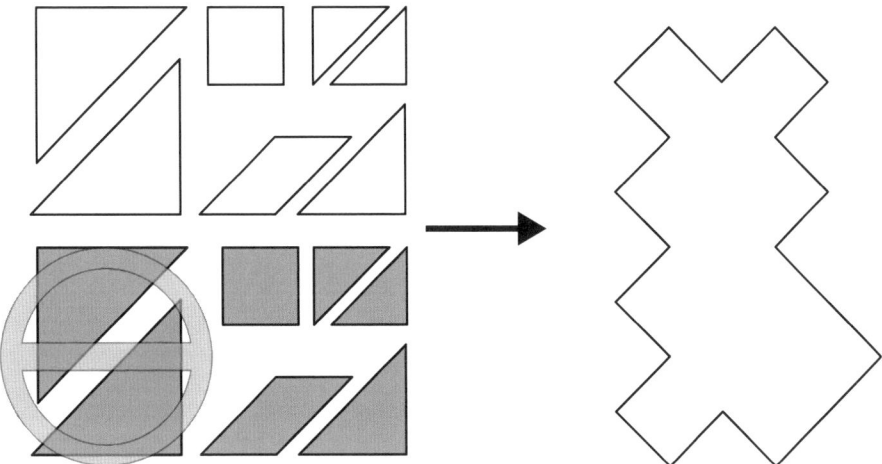

Step A: Arrange the given pieces to form the shape shown above.
Step B: Divide the shape into two identical parts.

Based on a puzzle by **Martin Gardner**, USA.

4.11. *The Chipped Triangle*

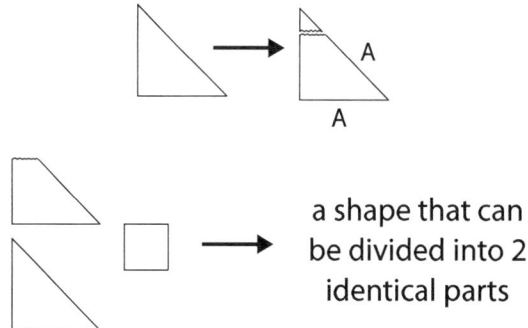

A large triangle was chipped so that the long edge is equal to the short edge (both edges are marked with an A).

Add a square and another large triangle to create a shape that can be divided into two identical parts.

This is the Tangram version of the Cuttrick puzzle by **Vesa Timonen**, Finland.

4.12. *The Mitre Puzzle*

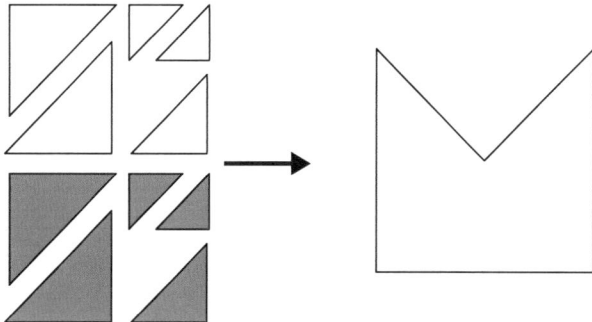

Step A: Arrange the given pieces to form the shape shown above.
Step B: Divide the shape into four identical parts.

In this puzzle, a "part" has no restrictions.

This is the Tangram version of the Mitre puzzle by **Sam Loyd**, USA.

4.13. *Many Folds, Single Cut*

Fold a square sheet of paper so that with a **single continuous cut**, the sheet of paper can be divided into the seven pieces of the Tangram.

This can be done if the sheet of paper is folded such that all the edges of the pieces (lines in the diagram above) are aligned.

Solutions

4.1. *The 49/50 Paradox*

The extra one unit of area ("dot") comes from the small spaces between the dots and the boundary or perimeter of the Tangram pieces.

4.2. *Contour Tangram Puzzles*

(a)

(b)

(c)

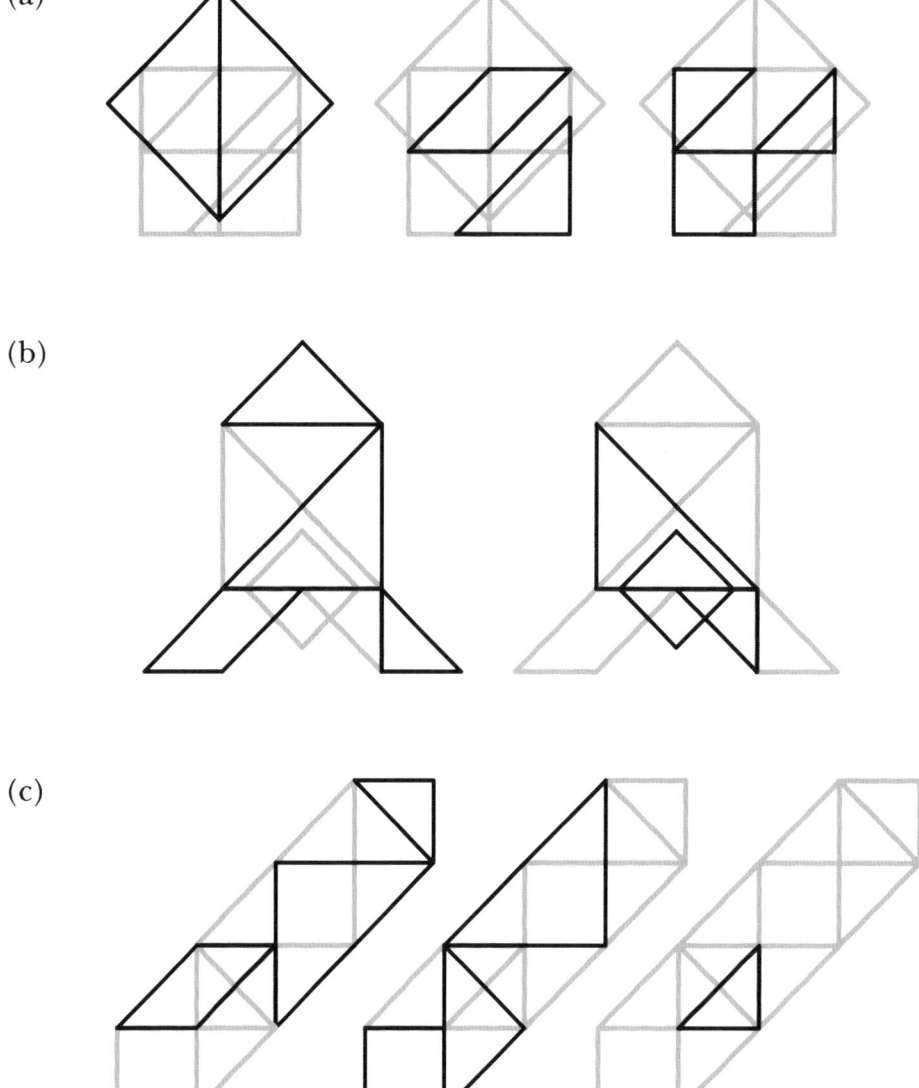

(d)

(e)

(f)

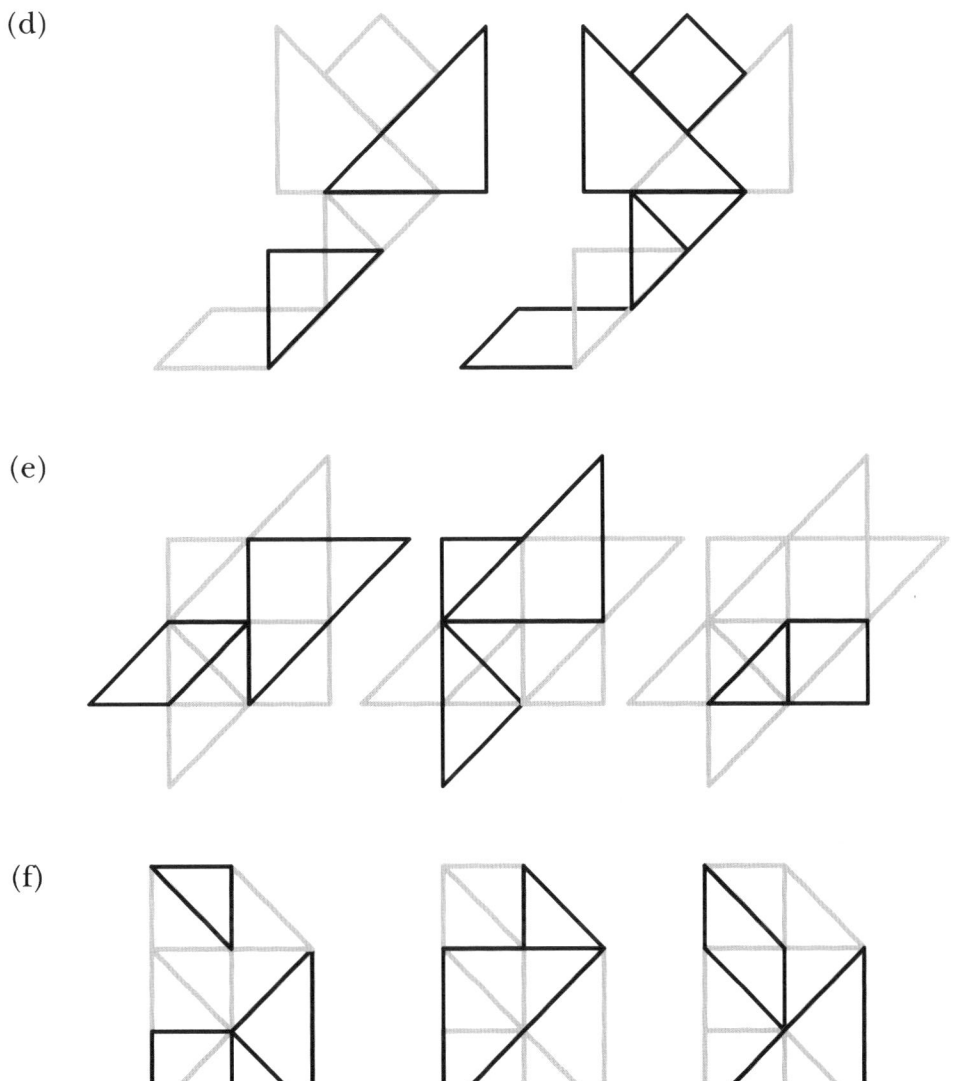

4.3. *Tangram Battleship*

Fleet 1

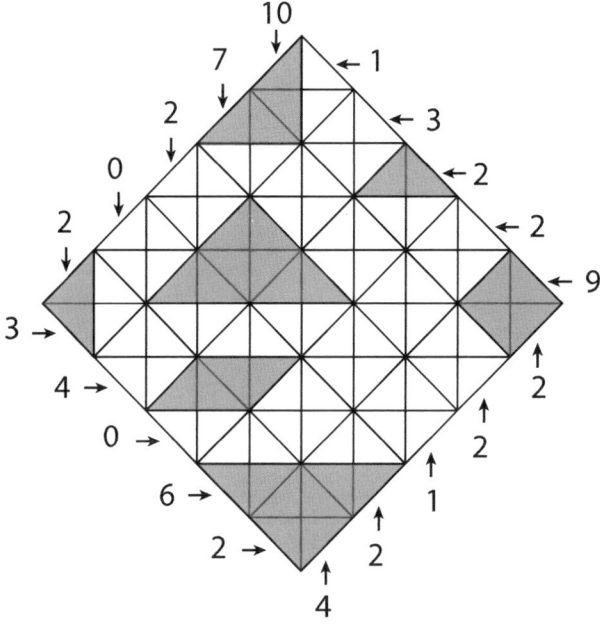

Fleet 2

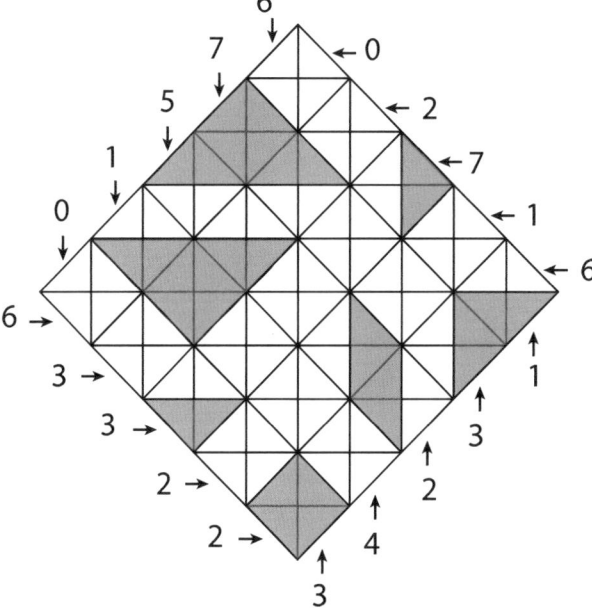

Fleet 3

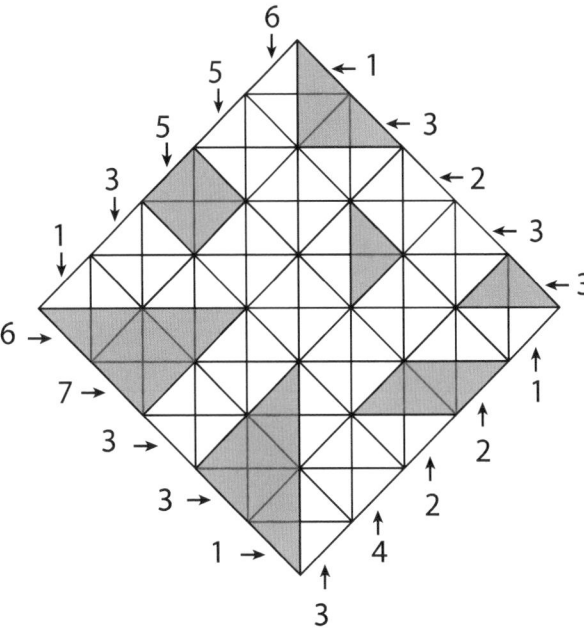

4.4. *TanFrames: Triangles*

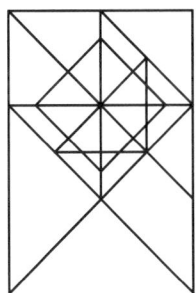

58 triangles of any size

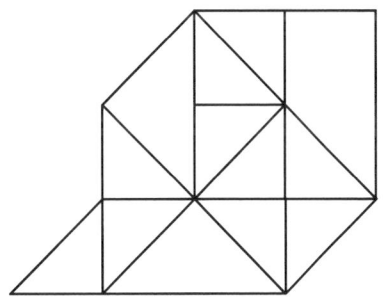

22 triangles of unit size

There are 58 possible triangular frames if there is no limit to the size of the triangles (left), and only 22 possible triangular frames if the size of the triangle is limited to a unit size (right).

4.5.   *30:60:90 Tangram*

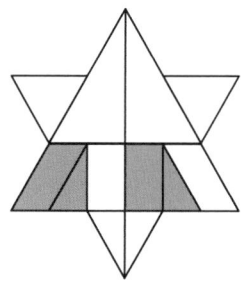

4.6.   *30:60:90 Silhouettes*

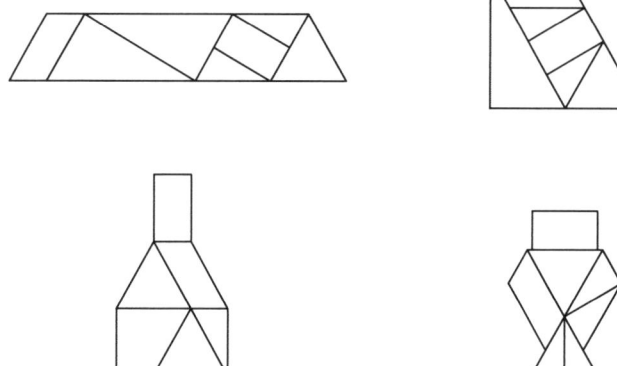

4.7.   *30:60:90 Stacking Trapeziums*

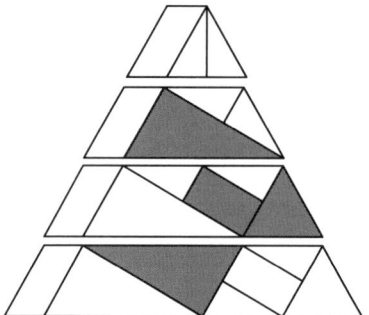

Take out the dark pieces to move up to the next level.

4.8. *The Matching Cube*

Step A

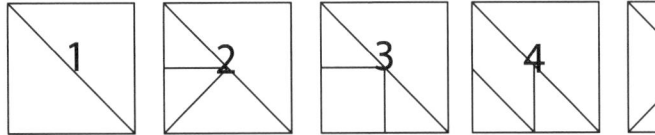

Step B

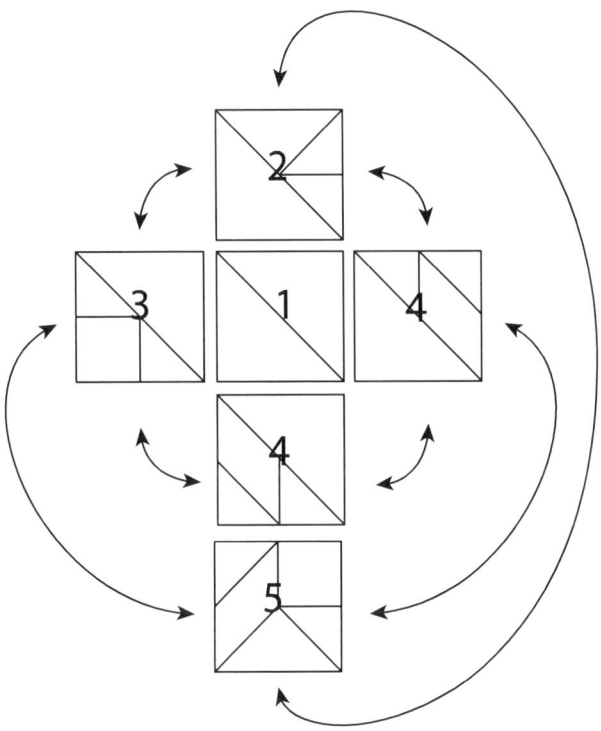

4.9. *The Square Skata*

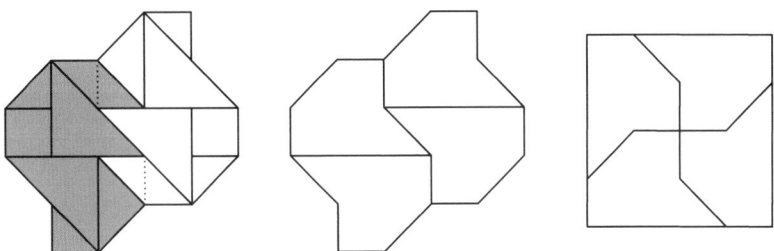

Cut the parallelograms along the dotted lines to create the four identical parts.

4.10.   *The Fishy Division*

Cut the gray parallelogram along the dotted line to create the two identical parts.

4.11.   *The Chipped Triangle*

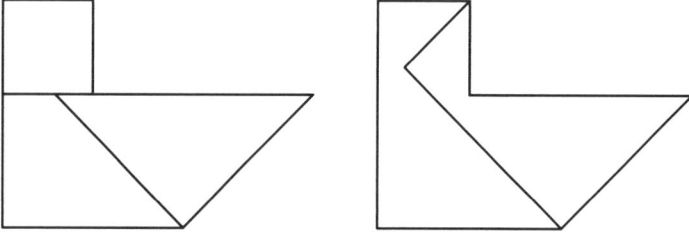

4.12.   *The Mitre Puzzle*

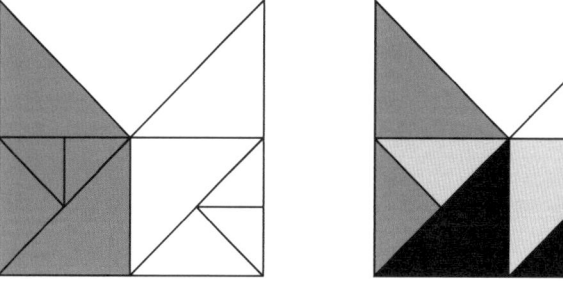

4.13.   *Many Folds, Single Cut*

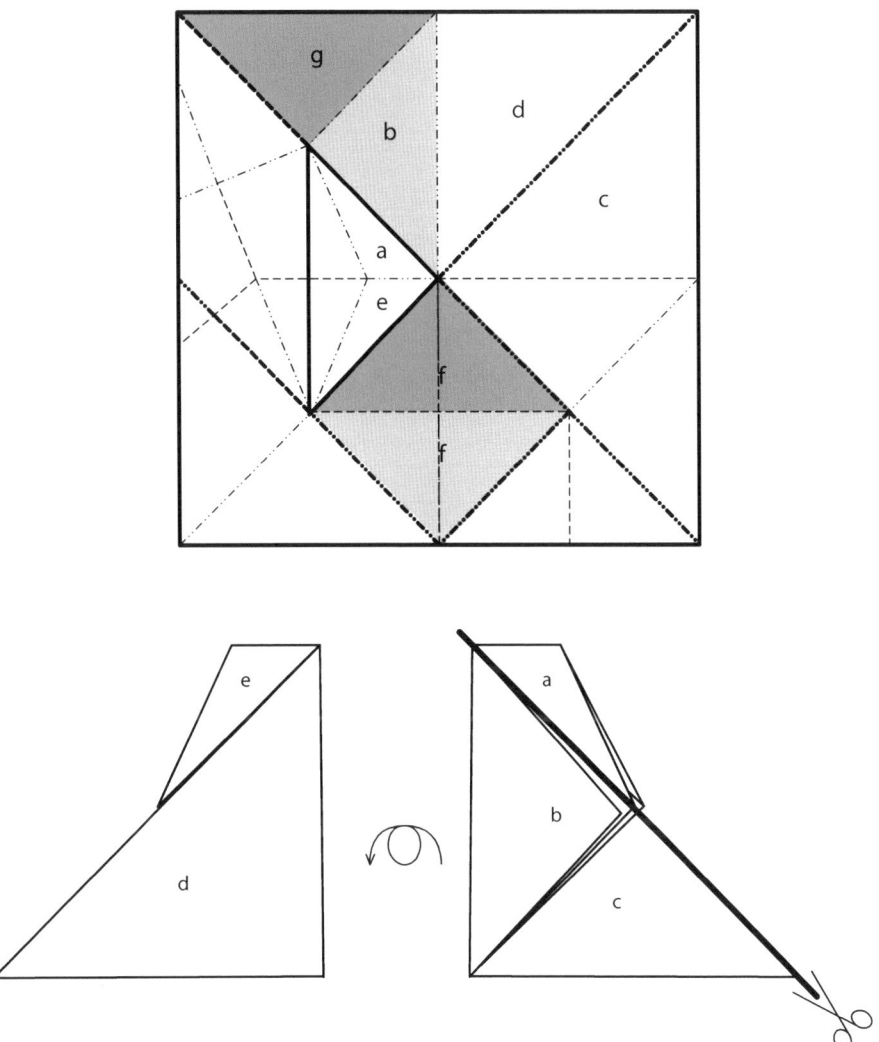

This is the crease pattern of the model. Print it out and fold along the lines.
Dashed lines are valley folds.

Dashed-dotted lines are mountain folds.

Thick lines are cutting lines.

The (f) triangles must be together when the paper is folded, while the
(g) triangle goes under the (b) triangle.

Appendix

A.1. *Triboloes*

Triboloes are shapes that are created with three right-angled isosceles triangles.

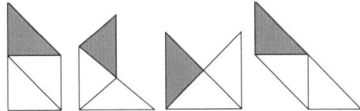

A.2. *Tetraboloes*

Tetraboloes are shapes that are created with four right-angled isosceles triangles. There are altogether 14 tetraboloes.

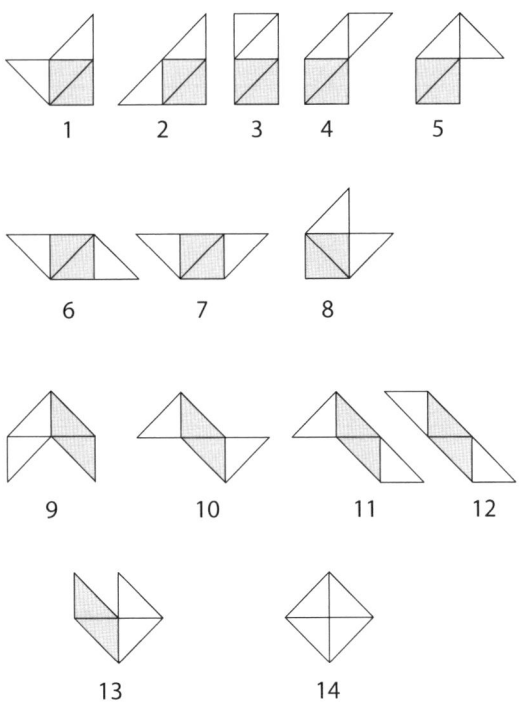

A.3. *Pentaboloes*

Pentaboloes are shapes that are created with five right-angled isosceles triangles. There are altogether 30 pentaboloes.

A.4.   *Pentominoes*

Pentominoes are shapes that are created with five squares. Each square must touch another along its full edge only. There are altogether 12 pentominoes.

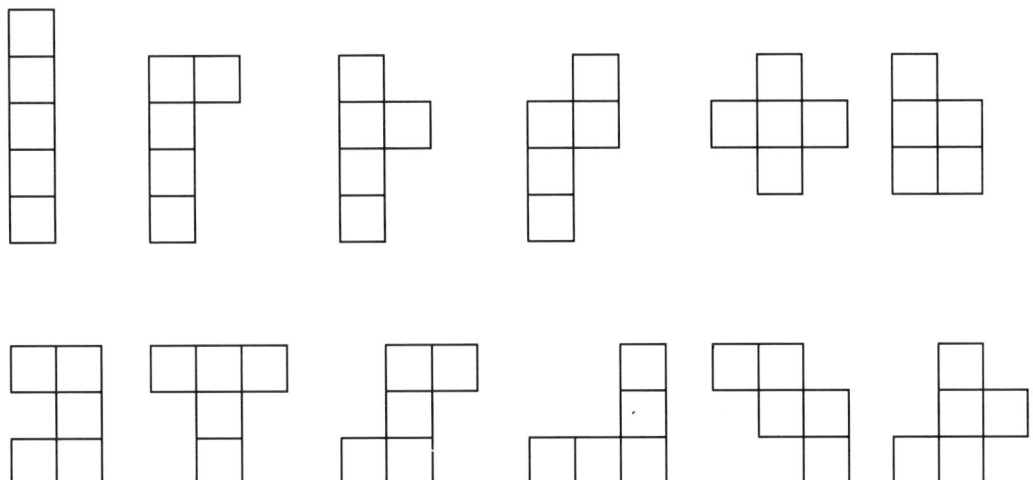